AF333879

Nanoscale Phenomena

Basic Science to Device Applications

LECTURE NOTES IN NANOSCALE SCIENCE AND TECHNOLOGY

Series Editors:

Zhiming M. Wang, Department of Physics, University of Arkansas, Fayetteville, AR, USA

Andreas Waag, Institut für Halbleitertechnik, TU Braunschweig, Braunschweig, Germany

Gregory Salamo, Department of Physics, University of Arkansas, Fayetteville, AR, USA

Naoki Kishimoto, Quantum Beam Center, National Institute for Materials Science, Tsukuba, Ibaraki, Japan

Volumes Published in this Series:
Volume 1: Self-Assembled Quantum Dots, Wang, Z.M., 2008

Volume 2: Nanoscale Phenomena: Basic Science to Device Applications, Tang, Z., and Sheng, P., 2008

Forthcoming Titles:

Volume 3: One-Dimensional Nanostructures, Wang, Z.M., 2008

Volume 4: Epitaxial Semiconductor Nanostructures, Wang, Z.M., and Salamo, G., 2008

Volume 5: B-C-N Nanotubes and Related Nanostructures, Yap, Y.K., 2008

Volume 6: Towards Functional Nanomaterials, Wang, Z.M., 2008

Nanoscale Phenomena

Basic Science to Device Applications

Zikang Tang and Ping Sheng

Editors

Hong Kong University of Science and Technology
Clear Water Bay, Hong Kong

 Springer

Zikang Tang
Department of Physics and Institute
of Nano Science and Technology
Hong Kong University
of Science & Technology
Clear Water Bay
Kowloon, Hong Kong

Ping Sheng
Department of Physics and Institute
of Nano Science and Technology
Hong Kong University
of Science and Technology
Clear Water Bay
Kowloon, Hong Kong

ISBN-13: 978-0-387-73047-9   e-ISBN-13: 978-0-387-73048-6

Library of Congress Control Number: 2007931614

Printed on acid-free paper.

9 8 7 6 5 4 3 2 1

springer.com

Preface

The Third International Workshop of the Croucher Advanced Study Institute (ASI) on Nano Science and Technology: from Basic Science to Device Applications, was held at Hong Kong University of Science and Technology from 8 January to 12 January 2007. The first and second workshops took place in January 1999, and January 2002, respectively.

Collected in this volume are 20 articles, 16 from invited talks and four from contributed presentations. The speakers are from the United States, Europe, Japan, Korea, Chinese Mainland, Taiwan, and Hong Kong. During the workshop, the vivid presentations captured the audience's attention not only with the great potential of nanotechnology, but also brought out the relevant underlying science. In Prof. Supriyo Datta's talk, a rather unique "bottom-up" view of electrical conduction was presented which is particularly relevant to nanoscale devices. He raised several thoughtful questions on transport physics such as dissipation and entanglement that could be operative in these devices. Professor Morinobu Endo described the advanced technology of large-scale synthesis of nanostructured carbons, with emphasis on novel applications that have already had an impact on our daily lives. His presentation impressed on us that nanotechnology applications are already here, with far-reaching implications ahead. Professor Ming-Chou Lin described his group's successful integration of silicon-based nano devices into biomolecular technology. With this integrated technology, the potential of silicon nanowires as a detector for DNA hybridization is demonstrated. Professor Vivian Yam delivered an excellent review of on her group's recent work on functional molecular materials. Through rational design and judicious functionalization and assembly strategies, she has demonstrated that many transition-metal molecular complexes possess structure-dependent light emitting characteristics with potential application as templates for the preparation of nano-sized materials. A rather unique presentation by Prof. Ping Sheng on nanoscale hydrodynamics showed that it is only recently that continuum hydrodynamics can quantitatively reproduce the nanoscale flow characteristics as simulated by molecular dynamics, and in so doing revises the traditional non-slip boundary condition. Taken together, these papers from the Third Croucher ASI gave a snapshot of the frontier research carried out over the past few years, and demonstrates that great progress has indeed been achieved both in fundamental research and in applications of nanoscience and technology.

We would like to express our gratitude to all of the authors for their excellent contributions. Special thanks go to Ms. Helen Lai and her team for the successful organization of this workshop, and to the Croucher Foundation for its financial support.

Zikang Tang and Ping Sheng
Kowloon, Hong Kong
May, 2007

Contents

Preface.. v

Part I   Nano Structured Carbons and their Applications

1   **Science and Technology of Nanocarbons**
Morinobu Endo... 3

2   **Catalytically-Grown Carbon Nanotubes and Their Current Applications**
Morinobu Endo... 9

3   **Transparent Conducting Films by Using Carbon Nanotubes**
Hong-Zhang Geng and Young Hee Lee.......................... 15

4   **Raman Spectroscopy on Double-Walled Carbon Nanotubes**
Wencai Ren and Hui-Ming Cheng.............................. 29

5   **Interface Design of Carbon Nano-Materials for Energy Storage**
Feng Li, Hong-Li Zhang, Chang Liu and Hui-Ming Cheng.......... 41

6   **Formation Mechanism of 0.4 nm Single-Walled Carbon Nanotubes in CoAPO-5 Single Crystals**
J. P. Zhai, I. L. Li, Z. M. Li, J. T. Ye and Z. K. Tang................ 49

Part II   Quantum Dots and Molecular Spintronics

7   **Nanodevices and Maxwell's Demon**
Supriyo Datta... 59

8   **Manipulating Electron Spins in an InGaAs/InAlAs Two-Dimensional Electron Gas**
C. L. Yang, X. D. Cui, S. Q. Shen, H. T. He, Lu Ding, J. N. Wang,
F. C. Zhang and W. K. Ge.. 83

9   Continuum Modelling of Nanoscale Hydrodynamics
Ping Sheng, Tiezheng Qian and Xiaoping Wang   99

**10   Defect in Zinc Oxide Nanostructures Synthesized by a
      Hydrothermal Method**
A. B. Djurišić, K. H. Tam, C. K. Cheung, Y. H. Leung, C. C. Ling,
C. D. Beling, S. Fung and W. K. Chan 117

Part III   Nano Materials Design and Synthesis

**11   Towards Surface Science Studies of Surfaces Formed by Molecular
      Assemblies Using Scanning Tunneling Microscopy**
Chen Wang and Shengbin Lei 133

12   Electronic Transport Through Metal Nanowire Contacts
Y. H. Lin, K. J. Lin, F. R. Chen, J. J. Kai and J. J. Lin 139

**13   Synthesis and Properties of Quasi-One-Dimensional Nitride
      Nanostructures**
Yong-Bing Tang, Dai-Ming Tang, Chang Liu, Hong-Tao Cong
and Hui-Ming Cheng 149

14   Electron Energy-Loss Spectroscopy for Nannomaterials
C. H. Chen and M. W. Chu 179

**15   Fabrication of Photovoltaic Devices by Layer-by-Layer
      Polyelectrolyte Deposition Method**
Wai Kin Chan, Ka Yan Man, Kai Wing Cheng and Chui Wan Tse 185

**16   Optical Properties of Arrays of Iodine Molecular Chains Formed
      Inside the Channels of $AlPO_4$-5 Zeolite Crystals**
J. T. Ye and Z. K. Tang 191

Part IV   Molecular Electronics

17   Quantum Manipulation at Molecule Scale
J. G. Hou 201

18   Silicon-Based Nano Devices for Detection of DNA Molecules
M. C. Lin, C. J. Chu, C. J. Liu, Y. P. Wu, C. S. Wu, and C. D. Chen,
I. C. Cheng, L. C. Tsai, H. Y. Lin, Ya-Na Wu, Dar-Bin Shieh, H. C. Wu
and Y. T. Chen 209

**19 From Simple Molecules to Molecular Functional Materials
      and Nanoscience**
Vivian Wing-Wah Yam and Keith Man-Chung Wong 217

Contentsix

20 First-Principles Method for Open Electronic Systems
Xiao Zheng and GuanHua Chen 235

Author Index 245

Subject Index 247

Contributors

C. D. Beling, Department of Physics, The University of Hong Kong, Pokfulam Road, Hong Kong

W. K. Chan, Department of Chemistry, The University of Hong Kong, Pokfulam Road, Hong Kong

Wai Kin Chan, Department of Chemistry, The University of Hong Kong, Pokfulam Road, Hong Kong

Cheng Hsuan Chen, Center for Condensed Matter Sciences, National Taiwan University, Taipei, Taiwan

Chii Dong Chen, Institute of Physics, Academia Sinica, Nankang 11529, Taipei, Taiwan

G. H. Chen, Department of Chemistry, The University of Hong Kong, Pokfulam Road, Hong Kong

X. D. Cui, Department of Physics, The University of Hong Kong, Hong Kong, China

Chen Wang, National Center for Nanoscience and Technology, Beijing, 100080, China

F. R. Chen, Department of Engineering and System Science, National Tsing Hua University, Hsinchu, 30013, Taiwan

Y. T. Chen, Department of Chemistry, National Taiwan University, Taipei 106, Taiwan

Hui-Ming Cheng, Shenyang National Laboratory for Materials Science, Institute of Metal Research, Chinese Academy of Sciences, Shenyang, 110016, China

Kai Wing Cheng, Department of Chemistry, The University of Hong Kong, Pokfulam Road, Hong Kong

I. C. Cheng, Department of Material Science, National University of Tainan, Tainan, 70005 Taiwan

C. K. Cheung, Department of Physics, The University of Hong Kong, Pokfulam Road, Hong Kong

M. W. Chu, Center for Condensed Matter Sciences, National Taiwan University, Taipei, Taiwan

C. J. Chu, Institute of Physics, Academia Sinica, Nankang 11529, Taipei, Taiwan

Hong-Tao Cong, Shenyang National Laboratory for Materials
Science, Institute of Metal Research,
Chinese Academy of Sciences,
Shenyang, 110016, China

Supriyo Datta, School of Electrical
and Computer Engineering,
Purdue University, West Lafayette, IN
47907, USA

Lu Ding, Department of Physics and
Institute of Nano-Science
and Technology, The Hong Kong
University of Science and Technology,
Hong Kong, China

A. B. Djurišiç, Department of Physics,
The University of Hong Kong,
Pokfulam Road, Hong Kong

Morinobu Endo, Faculty of
Engineering, Shinshu University,
4-17-1 Wakasato, Nagano-shi
380–8553, Japan

S. Fung, Department of Physics, The
University of Hong Kong,
Pokfulam Road, Hong Kong

W. K. Ge, Department of Physics
and Institute of Nano-Science and
Technology, The Hong Kong University
of Science and Technology,
Hong Kong, China

Hong-Zhang Geng, Department of
Nanoscience and Nanotechnology,
Department of Physics, and Center for
Nanotubes and Nanostructured
Composites, Sungkyunkwan Advanced
Institute of Nanotechnology,
Sungkyunkwan University, Suwon
440–746, R.O. Korea

H. T. He, Department of Physics and
Institute of Nano-Science
and Technology, The Hong Kong
University of Science and Technology,
Hong Kong, China

J. G. Hou, Hefei National Laboratory
for Physical Sciences at Microscale,
University of Science and Technology
of China, Hefei, Anhui 230026,
P.R. China

J. J. Kai, Department of Engineering
and System Science, National Tsing
Hua University, Hsinchu 30013, Taiwan

Young Hee Lee, Department of
Nanoscience and Nanotechnology,
Department of Physics, and Center for
Nanotubes and Nanostructured
Composites, Sungkyunkwan Advanced
Institute of Nanotechnology,
Sungkyunkwan University, Suwon
440–746, R.O. Korea

Y. H. Leung, Department of Chemistry,
The University of Hong Kong,
Pokfulam Road, Hong Kong

Feng Li Shenyang, National
Laboratory for Materials Science,
Institute of Metal Research, Chinese
Academy of Sciences, Shenyang
110016, China

I. L. Li, School of Engineering and
Technology, Shenzhen
University, Shenzhen, China

Z. M. Li, Department of Physics,
Hong Kong University of Science and
Technology, Clear Water Bay, Kowloon,
Hong Kong, China

Y. H. Lin, Institute of Physics, National
Chiao Tung University, Hsinchu 30010,
Taiwan

J. J. Lin, Institute of Physics & Department of Electrophysics, National Chiao Tung University, 1001 Ta Hsueh Road, Hsinchu 30010, Taiwan

Ming-Chou Lin, Institute of Physics, Academia Sinica, Nankang 11529, Taipei, Taiwan

K. J. Lin, Department of Engineering and System Science, National Tsing Hua University, Hsinchu 30013, Taiwan

H. Y. Lin Department of Molecular Science and Engineering, National Taipei University of Technology

C. C. Ling, Department of Physics, The University of Hong Kong, Pokfulam Road, Hong Kong

C. J. Liu, Institute of Physics, Academia Sinica, Nankang 11529, Taipei, Taiwan

Chang Liu, Shenyang National Laboratory for Materials Science, Institute of Metal Research, Chinese Academy of Sciences, Shenyang 110016, China

Ka Yan Man, Department of Chemistry, The University of Hong Kong, Pokfulam Road, Hong Kong

T. Qian, Department of Mathematics, The Hong Kong University of Science and Technology, Clear Water Bay, Kowloon, Hong Kong, China

Wencai Ren, Shenyang National Laboratory for Materials Science, Institute of Metal Research, Chinese Academy of Sciences, Shenyang 110016, China

S. Q. Shen, Department of Physics, The University of Hong Kong, Hong Kong, China

Ping Sheng, Department of Physics and Institute of Nano Science & Technology, Hong Kong University of Science and Technology, Clear Water Bay, Kowloon, Hong Kong China

Shengbin Lei, National Center for Nanoscience and Technology Beijing, 100080, China

Dar-Bin Shieh, Institute of Oral Medicine and Institute of Molecular Medicine, National Cheng-Kung University

Dai-Ming Tang, Shenyang National Laboratory for Materials Science, Institute of Metal Research, Chinese Academy of Sciences Shenyang 110016, China

K. H. Tam Department of Physics, The University of Hong Kong, Pokfulam Road, Hong Kong

Z. K. Tang, Department of Physics, Hong Kong University of Science and Technology, Clear Water Bay, Kowloon, Hong Kong, China

Y.-B. Tang, Shenyang National Laboratory for Materials Science, Institute of Metal Research, Chinese Academy of Sciences, Shenyang 110016, China

L. C. Tsai, Department of Material Science, National University of Tainan, Tainan 70005, Taiwan

Chui Wan Tse, Department of Chemistry, The University of Hong Kong, Pokfulam Road, Hong Kong

Chapter 1
Science and Technology of Nanocarbons

Morinobu Endo[*]

1.1 Introduction

Nanocarbons can be defined as carbon materials built from sp^2 bond building block in a nanoscale, and include various forms of carbons in the range from fullerene, carbon nanotube to nano-porous materials. Among them, one-dimensional carbon nanotubes have attracted a wide range of scientists due to their unique morphology and nano-sized scale, furthermore their versatile expected applications, in which the challenge is that how to utilize their intrinsic properties of carbon nanotubes, such as mechanical, thermal and electrical properties. It is no doubt that the recent explosive carbon research is due to the discovery of nano-scale carbon materials. After the identification of fullerenes and carbon nanotubes as new allotropes of carbon element in the end of the 20th century, chemists, physicists, scientists and engineers started to work intensively due to their extraordinary mechanical and electronic properties. In general, carbon is a unique element that has three types of bonding nature: sp^3, sp^2 and sp bonds, which result in diverse allotropes such as diamond, graphite, carbyne and fullerene. So, carbon differs chemically from Si, Ge, and Sn (the other group IV elements, which all have sp^3 bonding). Carbon has plenty of potential since it is the lightest atom in group IV of the periodic table. Now it is becoming the main element instead of silicon for electronic applications in this era, as Professor H. W. Kroto predicted, "21st century is the carbon age". The 19th century can be remembered as an iron age while the 20th century is recognized to be the century founded on silicon technology. Now much attention is paid on carbon "C" as one of the promising candidates and leading element in the 21st century. To date, carbons have contributed to welfares of human being as a "silent force behind the science". This is greatly due to the diverse structural forms and functions of carbons. Inherent advantage of carbons materials, originating from the extraordinary ability of the chemical element carbon to combine itself and other chemical elements in different ways, is blooming in the 21st century. Tailoring of carbon structure, that is, controlling the physicochemical properties of carbon materials on the nanometer scale, will be core technology for obtaining novel carbons

---

* Faculty of Engineering, Shinshu University, 4-17-1 Wakasato, Nagano-shi 380–8553, Japan

Z. Tang, P. Sheng (eds), *Nano Science and Technology.*
© Springer 2008

with new and extraordinary functions. The function and the added value of novel carbons based on nanotechnology will provide the industry of our society with a great business opportunity. Day by day, the request for revitalizing the global economy is growing. The only way to answer for this request is the technological innovation. According to model of Kondratieff the economist, new science will create a new industry and technology, and as a result, it will bring the wave of big economic activity. The generalization of science, technology and products in the 20th century, which led to the recession of world market, is said to be one of the reasons to the present economy slump. Now it is the time for new carbons based on carbon science to give rise to the creation of new functional materials, and finally technological innovation. As literally one of the major newspaper reported, "Carbon is hot now". Many scientists are being "hot" on carbon materials. The author and co-workers have been studying for many years how to control the nano-structure in order to enhance the function and performance, and develop new functions of carbon materials. Our ultimate goal is to be a "carbon tailor" that creates novel carbon structure by making use of sp^3, sp^2 and sp or hexagonal carbon structure to meet our need just like an experienced tailor making a made-to-order outfit. In this account, new carbons and their physical properties and novel functions from the viewpoint of nano-structure and its control will be presented with an emphasis on my own works.

1.2 Carbon Nanotubes

Nano-sized carbon nanotubes with hollow core were observed when hydrocarbons were decomposed in the existence of nano-sized catalyst such as iron at higher temperature (Oberlin et al., 1976). Interestingly, vapor-grown carbon fibers (VGCFs) shows very unique morphology such as an annular ring texture, and also contain single walled carbon nanotube or multi-walled carbon nanotube in the core part of the fibers (Fig. 1.1). Through a precise control of the synthetic conditions such as size of metal particles, it is possible to tailor the diameter, crystallinity and also the orientation of the cone angle with regard to the tube axis (Endo et al., 2001a; Terrones et al., 2001). As compared with those of arc discharge or laser vaporization,

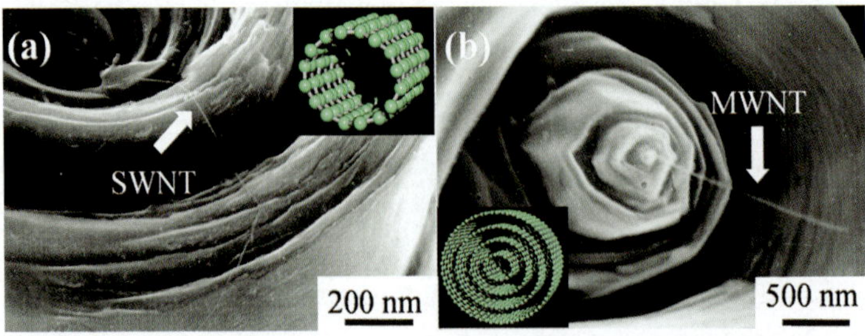

Fig. 1.1 FESEM images of vapour grown carbon fibers containing (a) SWNT and (b) MWNT

synthesis of carbon nanotube via a catalytic chemical vapor deposition (CVD) method has shown to be more controllable and profitable for the large-scale production. As a result, the diversity of microstructure and morphology of one-dimensional carbons are expected to be useful in various fields such as electronic devices, composites, biomaterials, batteries etc.

1.3 Anode Material for Lihitum Ion Battery

Since commercialization of Li-ion batteries (LIB) by Sony in 1992, they have been used in a wide range of mobile electronic devices, and now became leading product of secondary batteries (five hundred million battery cells per year). It is well known that carbon or graphite materials have been used as a main component of anode electrode in LIB. Actually, the advance of carbon science has been supporting the tremendous improvement of capacity of LIB for past 10 years (10% per year). These days, the new generation cellular phones having diverse functions with high data transfer rate require further enhancement of the battery performance. When considering the diverse and still growing LIB, e.g. from new type of cellular phone, high performance notebook, PC (personal computer), to power supply of 42V-vehicle and hybrid vehicle, it is critical to develop a higher performance LIB with advanced carbon materials as anode. In this sense, the outstanding mechanical properties and the high surface to volume ratio (due to their small diameter) enable carbon nanotubes to be incorporated as an additive in the commercial lithium-ion battery systems (Endo et al., 2001b). Astonishingly, only by adding small amount of high purity carbon nanotubes to the synthetic graphite, the cyclic efficiency was sustained at almost 100% up to 50 cycles. At higher concentrations, the nanotubes interconnect graphite powder particles together to form a continuous conductive network.

1.4 Electrode Material for Electric Double Layer Capacitor

Electric double layer capacitor (EDLC) has been expected as the electric energy storage device of the next generation, such as hybrid electric vehicle (HEV), fuel-cell vehicle, and power back-up system. Since the new interpretation of the pore structure is required in order to understand the mechanism of EDLC, we have developed an effective method for analyzing the nanopore with the help of transmission electron microscope (TEM) observations and image processing. For capacitance on EDLC, the theory for the conventional parallel plate condenser has been applied. The structure of the pore has been assumed to be the ideal structure such as slit and tube etc. However, actual pores that are quite different from these ideal models should be determined according to the TEM observations and image analysis process. The result of novel method we have developed is consistent with that of conventional gas adsorption method (Endo et al., 2001c,d,e). Furthermore, it is very helpful to understand the pore structure because the novel method analyzes the pore

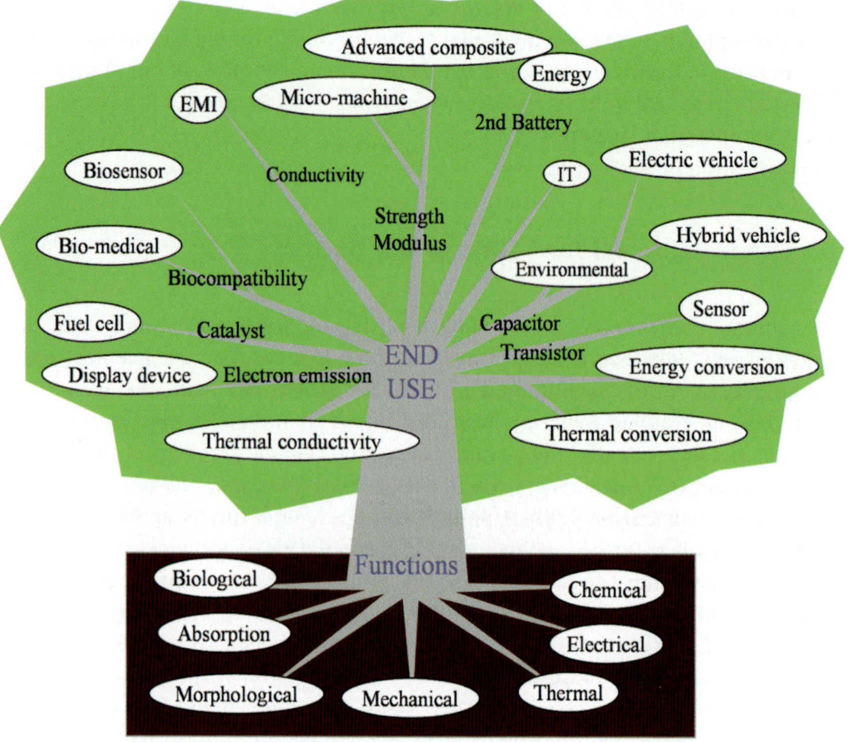

Fig. 1.2 Tree of nanocarbons with basic properties and applications

structure from different viewpoint. The novel method will be used as a powerful tool to develop a new carbon electrode for EDLC with high specific capacitance. Consequently, novel method shows some clues to control the nano-structure of carbon, and to get a final product with high performance. This enables us to effectively form an activated-carbon electrode with ideal pore size, which has the right room for the solvated ions in an aqueous and aprotic solvent system. All parameter in the process of preparing the capacitor (i.e. the choice of precursor, pyrolysis, activation etc.) should be considered for the end use. The creation of the new industry in relation with carbon based EDLC will be realized by the cooperation of carbon-related science and technology.

1.5 Conclusion and Prospect

It is estimated that the world market of nano materials is a twenty seven thousand billion yen market. This sum doubles when we incorporate the correlated industries and as a result, exceeds present automobile industry of this country. Novel carbons with new function and technology, constructed by tailor-made technology in nanoscale, will function as a starting point of industrial innovation. The technical development until the 20th century was mainly raised from the market needs. In the 21st century when there is no clear market needs, needs provoking-type technical innovation based on material technology is awaited. Applications fields of new carbons such as fuel cell, new type battery, composite with novel function, gas storage, and purification of water/air is considered to be the fundamental technology of the century. Conversion from conventional gasoline based vehicle to hybrid vehicle and fuel cell based vehicle, supported by new carbon technology, will cut the gas consumption to less than a half than usual. It has been expected that HEV will continue to increase its share around 10–20% of the domestic vehicles sales in 2007–2008, and at the same time, nano carbon industry is also expected to bloom. Widespread impact on nanocarbon-related technology on automobile industry and the housing field and promotion of the structural reform of the industry and economy is main target. Japanese consumers are renowned to be "green consumers" in the world, and this fact has much to do with the expansion of environmentally friend vehicle. Also, nano electron mechanical system will be developed by the contribution of nanocarbons as a sensor and actuator. New carbon as a key material of the 21st century will grow to a great industry and contribute to environmental, energy and IT both directly and indirectly. Intellectual production of nano carbon, that is, scientific results has been accumulated to a considerable amount, and it is time to the creation of the value (Fig. 1.2). When considering the recent research trend of carbons, we believe that this is not anymore a dream, and we are on the starting line.

References

Endo, M., Kim, Y.A., Fukai, Y., Hayashi, T., Terrones, M., Terrones, H., Dresselhaus, M.S., 2001a, Comparison study of semi-crystalline and highly crystalline multiwalled carbon nanotubes. *Applied Physics Letter*, **79**, pp. 1531–1533.

Endo, M., Kim, Y.A., Hayashi, T., Nishimura, K., Matushita, T., Miyashita, K., Dresselhaus, M.S., 2001b, Vapor-grown carbon fibers (VGCFs): Basic properties and their battery applications. *Carbon*, **39**, pp. 1287–1297.

Endo, M., Maeda, T., Takeda, T., Kim, Y.J., Koshiba, K., Hara, H., Dresselhaus, M.S., 2001c, Capacitance and pore-size distribution in aqueous and nonaqueous electrolytes using various activated carbon electrodes. *Journal of Electrochemical Society*, **148**, pp. A910–A914.

Endo, M., Kim, Y.J., Takeda, T., Hayashi, T., Koshiba, K., Hara, H., Dresselhaus, M.S., 2001d, Poly (vinylidene chloride)-based carbon as an electrode material for high power capacitors with an aqueous electrolyte. *Journal of Electrochemical Society*, **148**, pp. A1135–A1140.

Endo, M., Kim, Y.J., Maeda, T., Koshiba, K., Katayama, K., 2001e, Morphological effect on the electrochemical behavior of electric double-layer capacitors. *Journal of Material Research*, **16**, pp. 3402–3410.

Oberlin, A., Endo, M., Koyama, T., 1976, Filamentous growth of carbon through benzene decomposition. *Journal of Crystal Growth*, **32**, pp. 335–349.

Terrones, H., Hayashi, T., Munoz-Navia, M., Terrones, M., Kim, Y.A., Grobert, N., Kamalakaran, R., Dorantes-Davila, J., Escudero, R., Dresselhaus, M.S., Endo, M., 2001, Graphitic cones in palladium catalysed carbon nanofibers. *Chemical Physics Letter*, **343**, pp. 241–250.

Chapter 2
Catalytically-Grown Carbon Nanotubes and Their Current Applications

Morinobu Endo*

2.1 Introduction

One-dimensional carbon nanotubes have attracted a particular interest due to their unique morphology, nano-sized scale, novel physico-chemical properties and furthermore, their versatile applications (Dresselhaus et al., 1995; Iijima 1991; Oberlin et al., 1976; Saito et al., 1998). Carbon nanotubes could be visualized as rolled sheets of graphene with man-sized dimension (sp^2 carbon honeycomb lattice). They could be either single-walled with diameters as small as about 0.4 nm, or multi-walled consisting of nested tubes with outer diameters ranging from 5 to 100 nm. The constituent cylinders within multi-walled carbon nanotubes (MWNTs) may possess different chiral structures. In particular, the electronic properties of single-walled carbon nanotubes (SWNTs) vary from semiconductors to metals, depending upon the chiral angle (the way the hexagons are positioned with respect to the tube axis) (Dresselhaus et al., 1995; Saito et al., 1998).

In this chapter, we will describe large-scale synthetic methods for producing high-purity carbon nanotubes and their current usages in various fields from electrochemical system, multifunctional filler to medical devices. In addition, we will report the selective synthesis on highly pure and crystalline double walled carbon nanotubes (DWNTs) by the right combination of the catalytic chemical vapor deposition (CVD) method and the subsequent purification steps.

2.2 Large Scale Production of Carbon Nanotubes

In order to use carbon nanotubes in novel devices, it is necessary to produce these materials with a high crystallinity in large-scale at economic costs. In this context, the catalytic CVD method is considered to be the most optimum to produce bulk amounts of carbon nanotubes, particularly with the use of a floating catalyst method (Oberlin et al., 1976; Endo, 1988). This technique is more controllable and cost

---

* Faculty of Engineering, Shinshu University, 4-17-1 Wakasato, Nagano-shi 380-8553, Japan

Z. Tang, P. Sheng (eds), *Nano Science and Technology.*
© Springer 2008

efficient when compared to arc-discharge and laser ablation methods. Regarding the bulk production of nanotubes for industrial applications, it is important to mention that in the early 1990s, Hyperion Catalysis International, Inc. (Cambridge, MA) started the large-scale production of MWNTs. Other companies such as Applied Sciences, Inc and Showa Denko have already a large-scale production capacity for MWNTs, which exhibits relatively wide distribution of diameters ranging from 50 to 120 nm (avg. = 80 nm). The most interesting point is that all companies selected a catalytic CVD method, and furthermore three companies except Hyperion adopted the floating reactant technique for the large-scale production of MWNTs. The development of the floating reactant techniques allows a three-dimensional dispersion of the hydrocarbon together with the catalytic particles derived from the pyrolysis of organometallic compounds (Endo, 1988; Tibbetts et al., 1994; Kim et al., 2005). Therefore, it is clear that a great deal of progress has been achieved related to the bulk synthesis of carbon nanotubes. It is noteworthy that metallic compound in as-grown nanotubes, which is essential for nanotube growth but invoke some toxic properties, has to be removed as low as possible. In this sense, thermal treatment has been utilized as the powerful tool for reducing metal content below 100 ppm (Kim et al., 2005).

2.3 Selective Fabrication of Double Walled Carbon Nanotubes

A recently highlighted topic in carbon nanotube science and technology has been the synthesis and characterization of DWNTs. DWNTs with a co-axial structure, that occupies a position between MWNTs and SWNTs, have attracted much attention because they are expected to exhibit interesting structural and electronic properties. Recently we reported the fabrication of highly purified DWNTs (Endo et al., 2005a) through the right combination of a catalytic CVD method with an optimized two-step purification process. Interestingly, they have relatively homogeneous and small-sized inner (ca. 0.9 nm) and outer diameters (ca. 1.5 nm), they are thus expected to exhibit the 1D character of a quantum wire. By very carefully pouring a stable suspension of nanotubes into a PTFE filter-attached funnel, very thin (ca. 30 μm), round (diameter = 3.4 cm), light (weight = ca. 15 mg) and black bucky paper was produced (Fig. 2.1a). The resultant bucky paper is the result of the physical entanglement of DWNT bundles (10–30 nm) (Fig. 2.1b). From the cross-sectional HRTEM image (Fig. 2.1d), we observed that our tubes consist of two relatively round, small and homogeneous-sized (below 2 nm in the outer shell) two concentric individual tubules. Furthermore, these co-axial tubes are packed in a hexagonal array. This DWNT buckypaper is highly flexible and is mechanically strong enough to fold an origami. The high structural integrity in DWNT-derived bucky papers is thought to be derived mainly from their long lengths (up to μm scale), because the longer the tubes, the greater the mechanical robustness of the intermingled bundle.

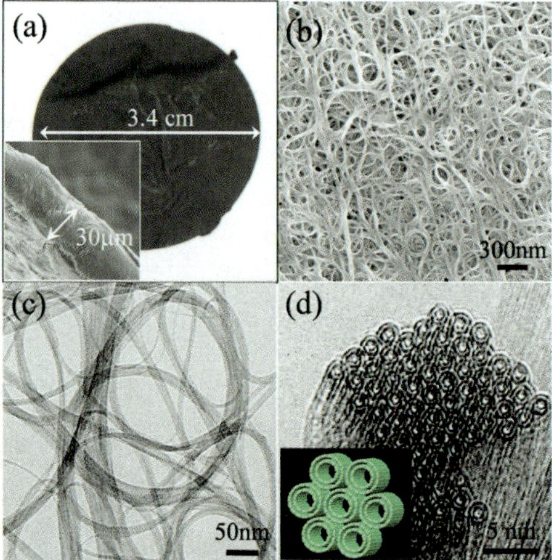

Fig. 2.1 (a) Photographs of round and thin DWNT-derived bucky paper (inset, thickness is ca. 30 µm). Low resolution transmission electron microscope (b) and field emission electron microscope (c) images indicate a bundled structure. (d) Cross-sectional high resolution transmission electron microscope image of a bundle of DWNT (inset is its schematic model, two concentric shells were regularly packed in a hexagonal array)

2.4 Novel Microcatheter from Carbon Nanotube-Filled Nylon

Among a wide range of potential applications from nano-composite, energy storage and energy conversion system, sensor, field emission displays and radiation source, nano-device, to probes, carbon nanotubes hold promise for medical application field. In this sense, the biocompatibility of carbon nanotubes has been actively investigated over the past few years. Recently, we fabricated carbon nanotube-filled nanocomposite-derived microcatheter via a conventional extruding system. Highly small-sized and black novel microcatheter exhibited highly reduced thrombus formation in vivo and low blood coagulation in vitro (Fig. 2.2) (Endo et al., 2001, 2005a,b). In addition, handleability of this novel microcatheter was enhanced due to increased elastic modulus (ca. 1.5 times) intrinsically caused by carbon nanotube-reinforcing effect. Novel microcatheter from nanotube-filled nanocomposite derived microcatheter exhibited good biological properties (e.g., reduced thrombus formation and low inflammatory reaction) that make them applicable in various fields from hemodialysis, cardiopulmonary bypass, intra-vascular catheters, left ventricular support to vascular dilating devices. Furthermore, our stable (almost inert) biological results based on the systematic T-cells study of high purity carbon nanotube clarified the potential toxic problem, and resultantly, will contribute to the active research on application of carbon nanotubes including medical fields.

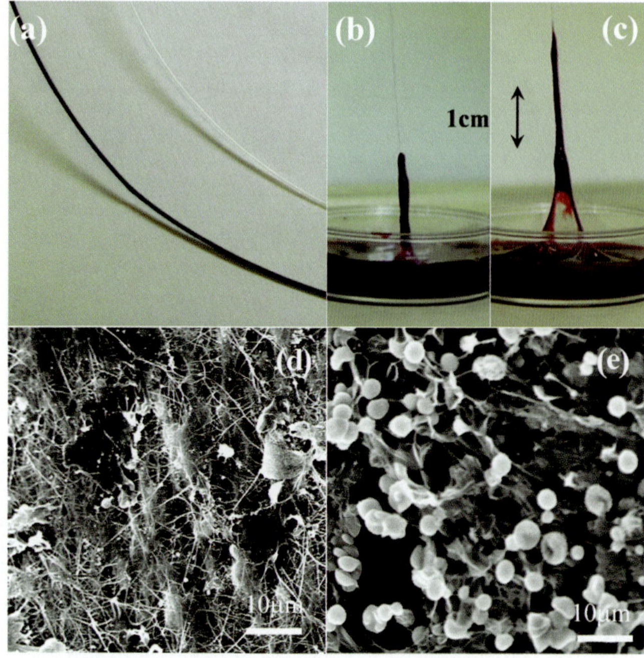

Fig. 2.2 (a) Photographs of transparent nylon-based and black nanotube-filled nanocomposite-derived microcatheter, blood coagulation of pure (b) and (c) nanotube-filled micro catheters, thrombus formation of (d) pure and (e) nanotube-filled micro catheters

2.5 Conclusion

This account has mainly described the current status and prospect regarding the large-scale synthesis of carbon nanotubes, with emphasis on their applications. The unique electronic properties of nanotubes make them also good candidates in the electronic-industry so that silicon-based technologies could be replaced by nano-carbons. At present, researchers are working on the following areas in order to apply carbon nanotubes in emerging technologies: (i) production of defect-free and high purity carbon nanotubes; (ii) establishment of useful techniques for quantifying the number of defects (types, where and amount etc.) in the nanotube structure; (iii) development of effective purification techniques (below ppm level) for the metal particles within carbon nanotube, especially for biological and electronic applications, and (iv) achieving homogeneous carbon nanotubes dispersions in polymer composites. The possibilities of applications using carbon nanotubes range from electronics, field emission display to energy storage devices and functional fillers in composites. Nowadays the carbon nanotubes-derived products have smeared into our life step by step, and before long, they will function as essential components for technological innovations.

References

Dresselhaus, M.S., Dresselhaus, G., Eklund, P.C., 1995, *Science of Fullerene and Carbon Nanotube* (Academic Press: Tokyo).

Endo, M., 1988, Grow carbon fibers in the vapour phase. *Chem. Tech.* **18**, pp. 568–576.

Endo, M., Saito, R., Dresselhaus, M.S., Dresselhaus, G., 1997, From carbon fibers to carbon nanotube. In *Carbon Nanotubes: Preparation and Properties*, edited by Ebbesen, T.W. (New York: CRC Press), pp. 35–110.

Endo, M., Kim, Y.A., Hayashi, T., Nishimura, K., Matushita, T., Miyashita, K., Dresselhaus, M.S., 2001, Vapor-grown carbon fibers (VGCFs): Basic properties and their battery applications. *Carbon*, **39**, pp. 1287–1297.

Endo, M., Muramatsu, H., Hayashi, T., Kim, Y.A., Terrones, M., Dresselhaus, M.S., 2005a, 'Buckypaper' from coaxial nanotubes. *Nature*, **433**, p. 476.

Endo, M., Koyama, S., Matsuda, Y., Hayashi, T., Kim, Y.A., 2005b, Thrombogenicity and blood coagulation of a microcatheter prepared from carbon nanotube-nylon-based composite. *Nanoletters*, **5**, pp. 101–106.

Iijima, S., 1991, Helical microtubules of graphitic carbon. *Nature*, **354**, pp. 56–58.

Kim, Y.A., Hayashi, T., Endo, M., Kaburagi, Y., Tsukada, T., Shan, J., Osato, K., Tsuruoka, S., 2005, Synthesis and structural characterization of thin multi-walled carbon nanotubes with a partially facetted cross section by a floating reactant method. *Carbon*, **43**, pp. 2243–2250.

Oberlin, A., Endo, M., Koyama, T., 1976, Filamentous growth of carbon through benzene decomposition. *Journal of Crystal Growth*, **32**, pp. 335–349.

Saito, R., Dresselhaus, G., Dresselhaus, M.S., 1998, *Physical Properties of Carbon Nanotubes* (Singapore: Imperial College Press).

Sun, X., Kiang, C.H., Endo, M., Takeuchi, K., Furuta, T., Dresselhaus, M.S., 1996, Stacking characteristics of graphene shells in carbon nanotubes. *Physical Review B*, **54**, pp. R12629–R12632.

Tibbetts, G.G., Bernnardo, C.A., Gorkiewicz, D.W., Alig, R.L., 1994, Role of sulfur in the production of carbon fibers in the vapor phase. *Carbon*, **32**, pp. 569–576.

Chapter 3
Transparent Conducting Films by Using Carbon Nanotubes

Hong-Zhang Geng and Young Hee Lee*

3.1 Introduction

The interests on the flexible transparent conducting films (TCFs) have been growing recently mainly because of strong needs for electrodes of flexible or wearable displays in the future (Cairns and Crawford, 2005). The current technology adopts indium-tin oxide (ITO) for TCFs that meets the requirement of low resistance and high transmittance. Yet, bending of the ITO film generates cracks in the film, resulting in poor flexibility reported by Saran et al. (2004). This is a serious drawback in flexible display applications. Carbon nanotube (CNT), which was discovered by Iijima (1991), is a new functional material that can be treated as graphitic sheets with a hexagonal lattice being wrapped up into a cylinder (Baughman et al., 2002). It has a high aspect ratio of typically 1,000 or greater with a diameter of a few nanometers. It is generally known that CNTs have high elastic modulus of 1–2 TPa as well as high electrical conductivity which is 1,000 times larger than Cu-wire (Bernholc et al., 1998; Salvetat et al., 1999; Mamalis et al., 2004). Wu et al. (2004) found that the CNTs were capable of forming naturally robust random network in the film and provide low sheet resistance and high transmittance with a minimal amount of CNTs. CNT films have been known to exhibit an excellent bending characteristics over the conventional ITO film. These structural and physical properties of CNTs are superb features for flexible TCFs, providing high performance of transparence and conductivity. Although CNTs have been suggested for flexible TCFs with outstanding film performance, the performance strongly relies on the film preparation conditions (Kaempgen et al., 2003; Sreekumar et al., 2003; Kim et al., 2003; Saran et al., 2004; Wu et al., 2004; Ferrer-Anglada et al., 2004; Hu et al., 2004; Zhang et al., 2005; Kaempgen et al., 2005; Moon et al., 2005; Unalan et al., 2006). The understanding for the film fabrication methods, material dependence, and criteria to determine the sheet resistance and transmittance is still in its infant stage.

* Department of Nanoscience and Nanotechnology, Department of Physics, and Center for Nanotubes and Nanostructured Composites, Sungkyunkwan Advanced Institute of Nanotechnology, Sungkyunkwan University, Suwon 440–746, R.O. Korea
e-mail: leeyoung@skku.edu

Z. Tang, P. Sheng (eds), *Nano Science and Technology.*
© Springer 2008

The fabrication of TCFs includes three steps: (i) the CNT treatment, (ii) the CNT-solution preparation, and (iii) the film preparation. The choice of CNTs, the degree of purity and defects of CNTs, and further post-treatment of CNTs are presumably important factors in determining the film performance. The CNTs are easily bundled with a diameter of typically a few tens nanometers due to their strong van der Waals interactions (~ 1000 eV) originating from the micrometer long CNTs (Thess et al., 1996; Pan et al., 1998; Sun et al., 2005). Thus, the macrodispersion (small-size bundles) or the nanodispersion (dispersion into individual nanotubes) is strongly desired in order to minimize the absorbance (O'Connell et al., 2001; Richard et al., 2003; Lee et al., 2005; Kim et al., 2005). The choice of solvents and the related dispersants are crucial factors in achieving the best dispersion conditions (O'Connell et al, 2002; Vaisman et al, 2006). Once the CNT solution is ready, the film preparation is another important step to consider, since this determines the networking of CNTs.

The purpose of this work is to determine which type of nanotubes is the best for flexible TCFs. Several single-walled CNTs (SWCNTs) prepared by different methods such as arc discharge (Arc), catalytic chemical vapor deposition (CVD), high pressure carbon monoxide (HiPCO), and laser ablation (Laser) were introduced to search for the material parameters. These CNTs were dispersed in water with sodium dodecyl sulfate (SDS) with sonication. This CNT solution was then sprayed onto the Poly(ethylene terephthalate) (PET) substrate by a spray coater to form the CNT film. The quality of CNTs, the degree of dispersion, film morphology and the performance of TCFs were characterized by scanning electron microscopy, transmission electron microscopy, thermogravimetric analysis, Raman spectroscopy, optical spectra, and four-point probe measurements. We found that the material parameters such as the diameter and crystallinity of CNTs, bundle size, networking of CNTs of the film, the degree of dispersion of CNTs in solvent are crucial parameters to improve the film performance. We proposed a material quality factor to evaluate the quality of SWCNTs and further discussed the film performance that relied on the absorbance and on the degree of entanglement in the CNT network. Our analysis demonstrates that the Arc SWCNTs are the best candidate to give rise to the lowest sheet resistance and highest transmittance among other species. The CNT film performance shows a good candidate for replacing the conventional ITO transparent conducting film in various application fields.

3.2 Experimental

Different kinds of SWCNT materials were used in this study. The CVD SWCNTs were purchased from SouthWest NanoTechnologies, Inc., HiPCO SWCNTs and Laser SWCNTs were from Carbon Nanotechnology, Inc. Arc SWCNTs were from Iljin Nanotech Co. Ltd. The CNTs were characterized by the field-emission scanning electron microscopy (FE-SEM-JSM7000F), transmission electron microscopy (TEM-JEM2100F), Raman spectroscopy (Renishaw RM1000-Invia), and thermogravimetric analysis (TGA-Seiko Exstar 6000 (TMA6100), SEICO INST. JAPAN).

Deionized water was used to disperse SWCNTs with sonication and centrifuge process. SWCNTs with a concentration of 0.3 mg/ml and SDS (Sigma–Aldrich) of 3 mg/ml were dissolved in water and sonicated in a bath type sonicator (Power sonic 505) at 400 W for 10 hours. The SWCNT solution was centrifuged at 10,000 g for 10 minutes by centrifuger (Hanil Science Industrial, MEGA 17R). The upper 50% of the supernatant solution was carefully decanted for characterization. Absorbance of the SWCNT supernatant was recorded by UV-VIS-NIR spectrophotometer (Cary-5000). The SWCNT supernatant was directly sprayed with air brush pistol (GUNPIECE GP-1, FUSO SEIKI Co. LTD) onto the PET substrate. A special care for the choice of CNT concentration was taken in order to keep the nozzle from being clogged.

During the spray process, the PET substrate was kept on the sample holder at 100 °C to accelerate evaporation of the fine droplets on the surface. When the spray process was terminated, the TCF film was immersed into deionized water for 10 minutes to remove the surfactant and dried in a dry oven at 80 °C for 30 minutes. This process was repeated twice to further remove surfactants and enhance adhesion between CNTs and PET film. The transmittance was recorded by UV-VIS-NIR spectrophotometer in the visible range (400–800 nm). Measurements of the sheet resistance were carried out by four-point probe method (Keithley 2000 multimeter) at room temperature.

3.3 Results and Discussion

3.3.1 Analysis on SWCNT Materials

The general properties of SWCNT powders were characterized by different methods. Figure 3.1 presents the FE-SEM and TEM morphologies of different SWCNT powders used in our study. The average diameters of individual CVD, HiPCO, Laser, and Arc SWCNTs were 0.8, 1.0, 1.2, and 1.4 nm, respectively, as determined from the TEM images. The CVD SWCNTs had the smallest average bundle size, as estimated from the SEM images, whereas the Laser sample had the largest average bundle size among all the samples. Some carbonaceous particles on the CNT bundles were present in the CVD SWCNTs. The Arc SWCNTs had relatively well-defined crystallinity without amorphous carbons on the tube walls, although the bundle size of the Laser sample was larger than that of the Arc sample from the FE-SEM images. The presence of carbonaceous particles on the nanotube walls is an important factor for the application of TCFs, since they determine the contact resistance among the CNT network (Dai et al., 1996; Jost et al., 1999; Hu et al., 2006).

The purity of each sample can be obtained from the TGA. Figure 3.2 presents the TGA and differential TGA of the SWCNTs that were used in this work. The SWC-NTs started to burn off in air with increasing temperature. The remaining material at 900 °C was metal oxides. All the samples contained some amounts of catalysts. The burning temperature obtained from the peak of differential TGA varied with several

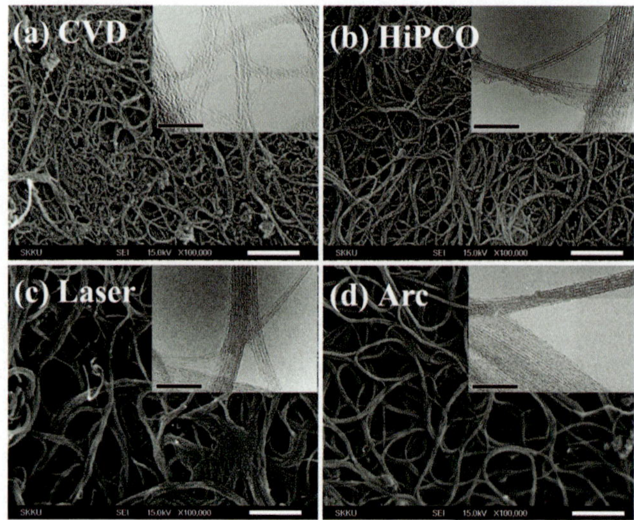

Fig. 3.1 FE-SEM (scale bar: 200 nm) and TEM (inset, scale bar: 20 nm) images of CNTs: (a) CVD SWCNTs, (b) HiPCO SWCNTs, (c) Laser SWCNTs, and (d) Arc SWCNTs

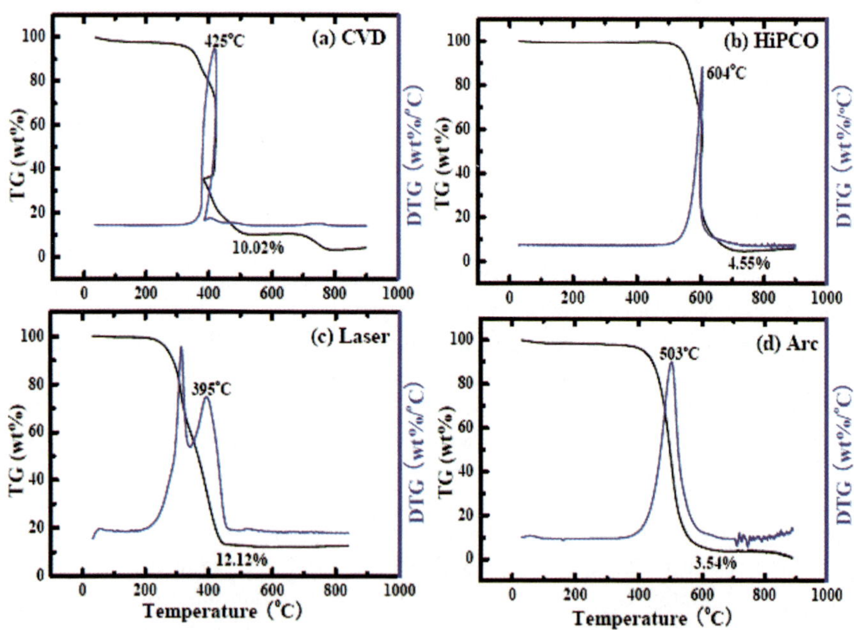

Fig. 3.2 TGA and differential TGA of the CNTs: (a) CVD SWCNTs, (b) HiPCO SWCNTs, (c) Laser SWCNTs, and (d) Arc SWCNTs

parameters, for instance, the content of metals, the bundle diameter of CNTs, the number of walls, and the crystallinity of CNTs (Arepalli et al., 2004). It is interesting to note the strong correlation between the sample purity (or catalyst amount) and the burning temperature for all the SWCNTs in Fig. 3.2a–d. The unusual endothermic burning behavior was observed at the sample typically with high metal content, which can be seen in the CVD SWCNTs in Fig. 3.2a. The Laser SWCNTs in Fig. 3.2c revealed two peaks in the differential TGA. The first peak could be identified as the burning temperature of amorphous carbons (Shi et al., 2000; Sadana et al., 2005). The second peak could be regarded as the SWCNT peak but the burning temperature was low mainly due to the presence of large amount of catalysts (~ 10 wt%). The burning temperature can be also related to the number of defects on the CNT walls.

The SWCNT powders were also characterized by Raman spectroscopy. Figure 3.3 shows the Raman shift of the samples at excitation energies of 514 nm (2.41 eV) and 633 nm (1.96 eV). The metallicity of each sample can be determined from the radial breathing modes (RBMs). For a given laser excitation energy, the SWCNTs whose allowed transition energies between van Hove singularities are in agreement with the excitation energy, will be excited to produce the Raman signal. This is known as the resonant Raman spectroscopy. Since the positions of van Hove singularities are dependent specifically on the diameter and chirality,

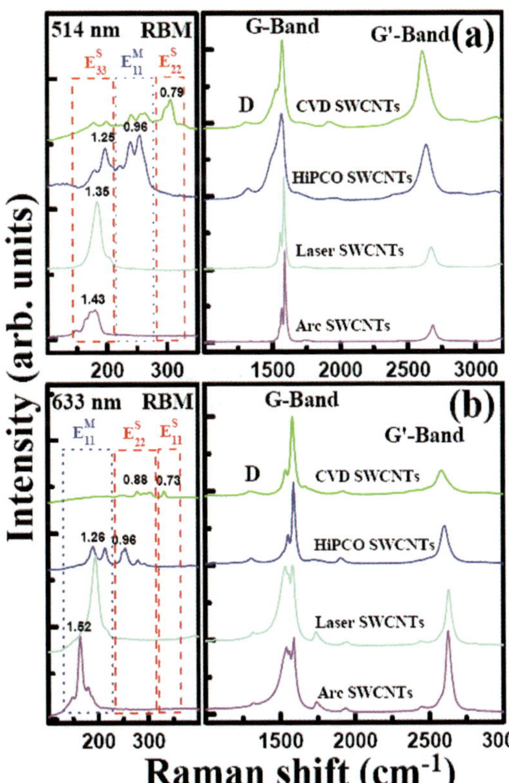

Fig. 3.3 Raman spectra of the SWCNT materials at excitation energy of (a) 514 nm and (b) 633 nm. Each region of van Hove singularities is provided in the box

the metallicity of the excited SWCNTs can be determined thoroughly (Rao et al., 1997; Jorio et al., 2001; Kuzmany et al., 2001). At 514 nm, the Laser and the Arc samples revealed only the semiconducting nanotubes. On the other hand, the CVD and the HiPCO samples contained both metallic and semiconducting nanotubes. The metallic component of SWCNTs was further evidenced by the presence of long tail at lower energy side in the G-band, i.e., the Fano line representing the metallic contribution (Brown et al., 2001a). At 633 nm, the Laser and the Arc samples picked up mostly metallic SWCNTs, whereas the CVD sample retained mostly semiconducting SWCNTs and the HiPCO sample contained both metallic and semiconducting SWCNTs. In addition to RBMs and G-band, the D-band was also present in all samples. D-band near $1320\,cm^{-1}$ originates in general from the presence of defects, although its exact origin was not certain (Brown et al., 2001b). The intensity of D-band indicates the amount of defects on the nanotube walls. Another intriguing parameter is the intensity of the G'-band. The G'-band is the second-order Raman signal, which is the first overtone of the D-band. Geng et al. (2007) found that G'-band is governed by the metallicity of the CNTs, i.e., the intensity of G'-band is directly proportional to the metallic composition of nanotubes. At 514 nm, the Arc and Laser samples revealed relatively small intensities, because they contained mostly semiconducting nanotubes, as shown in Fig. 3.3a. On the other hand, the CVD and HiPCO samples contained a certain portion of metallic composition, therefore yielding the large intensities of the G'-band. At 633 nm, similar trend of the changes in the intensity of G'-band was observed. The intensity of G'-band increased with increasing metallic composition in each sample.

3.3.2 Material Quality Factor for SWCNTs

We obtained several parameters of the SWCNTs used in our study. Figure 3.4 presents the summary of the material parameters of the SWCNT samples. The average diameters of each sample were determined from the TEM. Arc SWCNTs had the largest average diameter among other types, as shown in Fig. 3.4a. Figure 3.4b shows the burning temperature changing with the purity of SWCNTs. More impurities such as transition metals and carbonaceous particles indicate degradation of conductivity. Figure 3.4c presents the intensity ratio of D-band to G-band and G'-band to G-band at excitation energies of 514 and 633 nm, respectively. The presence of defects is represented in terms of the relative intensity ratio of D-band to G-band. This leads to lower the electrical conductivity of metallic CNTs. The abundance of metallic components is represented by the intensity ratio of G'-band to G-band. More metallic nanotubes results in higher conductivity of the materials.

The band gap of semiconducting nanotubes is inversely proportional to the diameter, $E_g = 2\,a_{c\text{-}c}\,\gamma_o/D$ (eV), where $a_{c\text{-}c}$ is 0.142 nm and γ_o is an empirical tight-binding parameter taken as 2.9 eV (Dresselhaus et al., 2005). The conductivity of SWCNTs can generally be simply expressed as

$$\sigma = ne\mu_n + pe\mu_p \tag{3.1}$$

Fig. 3.4 Summary of the material parameters for SWCNTs: (a) diameters, (b) burning temperature and purity, (c) The ratio of the intensity of D-band to G-band and G′-band to G-band of the SWCNTs at excitation energies of 514 and 633 nm

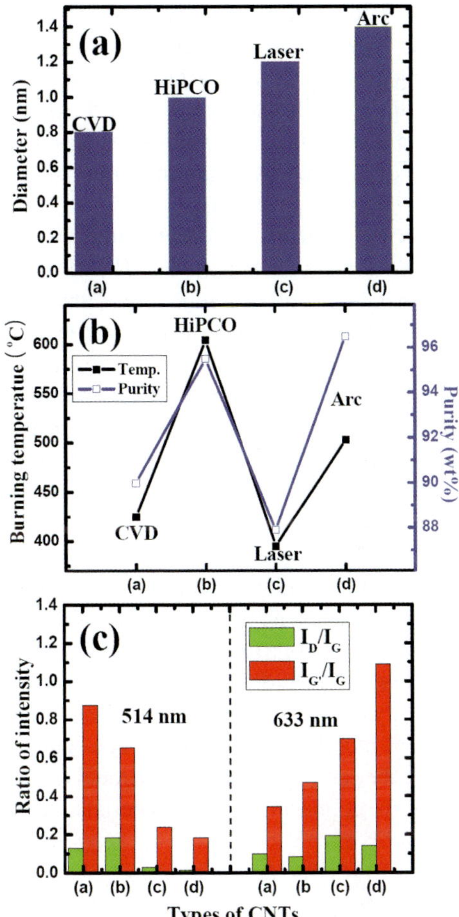

where n and p are the n-type (electrons) and p-type (holes) carrier concentrations, respectively, and μ_n and μ_p are the respective electron and hole mobility. The mobility is dominated by a succession of random scattering from collisions with lattice atoms, impurity atoms, and other scattering centers. The intrinsic carrier concentration decreases exponentially with band gap, $n_i = n_o \, exp(-E_g/2k_BT)$, where k_B and T are the Boltzmann constant and temperature of the system, respectively. The p-type and n-type nanotube carrier concentrations are, $p = n_i \, exp[(E_i - E_f)/k_BT]$, and $n = n_i \, exp[(E_f - E_i)/k_BT]$, where the intrinsic Fermi level E_i is frequently used as a reference level when the extrinsic semiconductors are discussed with a Fermi level of E_f. The conductivity is proportional to the carrier concentration, $\sigma \sim exp(-E_g/2k_BT)exp[(E_i - E_f)/k_BT]$ for semiconducting nanotubes. For intrinsic semiconducting nanotubes, $E_i = E_f$. On the other hand, in metallic nanotubes, π and π^* overlap at the Fermi level, i.e., metallic nanotubes are always metallic independent of the diameters. Nevertheless, Delaney et al. (1998) had suggested that the nanotubes are usually bundled and a pseudogap of ~ 0.1 eV is open due to the tube-tube interaction. This pseudogap is small compared to the direct bandgap of semiconductors with diameters of $1 \sim 1.4$ nm corresponding to bandgaps of $0.7 \sim 1.0$ eV (An and Lee, 2006). The pseudogap in metallic SWCNT bundles is inversely proportional to the tube diameter via $E_{pg} \approx 0.105/D$ (eV) after fitting to the observed values (Ouyang et al., 2001). Thus, the conductivity of metallic nanotubes reveals the similar diameter dependence to semiconducting ones, $\sigma \sim exp(-E_{pg}/2k_BT)$.

As we described before, the intensity of the G'-band is a measure of the abundance of metallic nanotubes. Therefore we propose this concept in quantifying the metallicity of nanotubes. It is also noted in our samples that the intensity of the G'-band is strongly correlated to the intensity of metallic peaks in RBMs. Despite the abundance of metallicity, the presence of defects on the nanotube walls that may act as scattering centers degrades the conductivity of the SWCNT network. The intensity of the D-band indicates the amount of defects on the nanotube walls. Therefore, an appropriate parameter to express conductivity of nanotubes for SWCNT films is the intensity ratio, G'-band/D-band. High abundance of metallicity and few defects on the nanotube walls will be desired for high conductivity of the SWCNT films.

From these analyses, it may be useful to extract a practical parameter as a figure of merit to evaluate the metallicity of SWCNTs. The purity affects the conductivity. The diameter contributes to the conductivity via bandgap described in the previous paragraph. More defects reduce the *mean free path* of carriers and decrease the mobility of carriers in nanotubes. The intensity ratio of G'-band/D-band may represent the mobility of the carriers. The conductivity is proportional to the metallicity of nanotubes and inversely proportional to the number of scattering centers or defects. Here, we define an effective material quality factor Q_m that governs the conductivity of SWCNTs

$$Q_m = P \times (e^{\frac{-E_g}{2k_BT}} \times e^{\frac{E_i - E_f}{k_BT}} \times \overline{\sum I_S} + e^{\frac{-E_{pg}}{2k_BT}} \times \overline{\sum I_M}) \tag{3.2}$$

where $E_g = 0.82/D$ (eV) and $E_{pg} = 0.105/D$ (eV) from the previous paragraph, D (nm) is the average diameter of individual SWCNTs and P is the purity of the sample. The intensity ratio of G'/D was averaged over the excitation energies of Raman spectroscopy for each metallic and semiconducting nanotubes. Here I_S (I_M) is defined as

$$I_S(I_M) = I_{G'/D} \times \frac{A_S(A_M)}{A_M + A_S} \tag{3.3}$$

where A_S (A_M) is the area intensity of semiconducting (metallic) peaks of RBMs from Raman shift in Fig. 3.3. The first exponential term in Eq. (3.2) represents the carrier concentration and the second exponential term represents the mobility that changes with doping effect from semiconducting nanotubes. The second term represents the contribution from metallic nanotubes. Thus this formula resembles the conductivity of Eq. (3.1). Only two wavelengths were used in our study but it

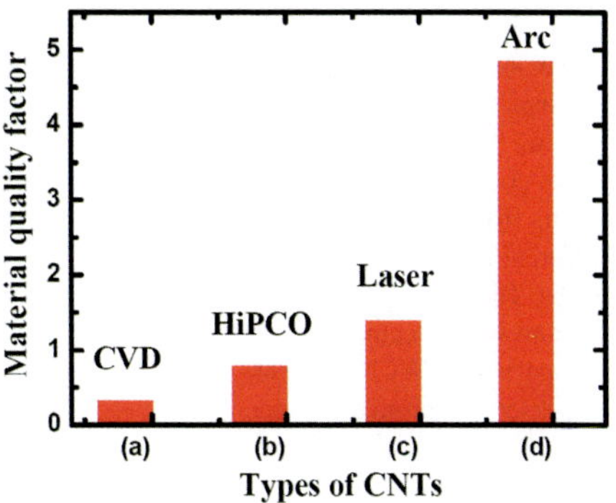

Fig. 3.5 The material quality factor of SWCNTs defined by Eq. (3.2) in the text

seemed satisfactory to explain the TCF performance in terms of Q_m. The abundance of metallic nanotubes is implicitly expressed in the intensity ratio of G'/D. Since the exponent of metallic nanotubes is much greater than that of semiconducting nanotubes with typical diameters of $1 \sim 2.0$ nm, the first term can be negligible in Eq. (3.2) in intrinsic nanotubes when $E_i \approx E_f$. The semiconducting nanotubes generally are sensitive to the environment invoking a doping effect. In such cases, the Fermi level can be shifted and the first term may not be negligible. Although the pseudogap of metallic nanotubes can be affected by a doping effect (Kwon et al., 1998; Zhu et al., 2005), we didn't consider it in Eq. (3.2), since it is relatively small in comparison to that of semiconducting nanotubes. We calculated the material quality factor Q_m in case of $E_i \approx E_f$. Figure 3.5 shows the material quality factor Q_m of the four SWCNTs in our study. From this point of view, the Arc sample was presumably the best sample that provided the highest conductivity than other types of SWCNTs considered in this work, as can be compared from Fig. 3.5.

3.3.3 Film Characteristics of SWCNTs

Once the prepared SWCNT powder was dispersed in deionized water with SDS, as described in experiment, this solution was sprayed onto the PET film to form thin SWCNT TCFs. Figure 3.6 presents the characteristic curves for sheet resistance-transmittance. For a direct comparison, at 80% of transmittance, the Arc TCFs showed the lowest sheet resistance of about 160 Ω/sq, whereas the CVD TCFs showed the highest sheet resistance. From the material point of view, as discussed in the previous paragraph, this general trend of the film performance seemed to be quite understandable.

We calculated the sheet conductance of TCFs at transmittance of 70 and 80% from Fig. 3.6 and plot as a function of material quality factor. Figure 3.7 shows the sheet conductance of TCFs at a transmittance of 70 and 80% as a function of the material quality factor Q_m. We clearly observed a direct correlation between the sheet conductance and the material quality factor. The sheet conductance reveals a linear relationship with the material quality factor. Although this empirical formula

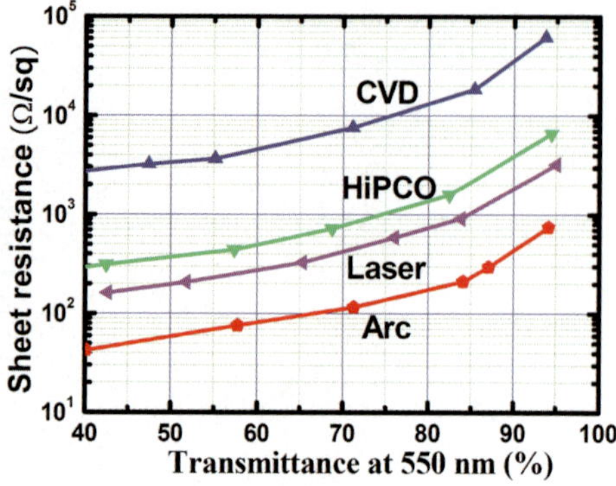

Fig. 3.6 Characteristic curves of sheet resistance-transmittance for various SWCNT TCFs. Each curve contains several data points from TCFs with different number of spray times. Transmittance at 550 nm was selected for comparison. SWCNTs were dispersed in deionized water with SDS

is not rigorous, it can provide at least a means of estimating material quality that governs the conductivity of the SWCNT TCFs. For instance, large diameter, higher purity, less defects (lower intensity of D-band), and more metallic nanotubes (higher intensity of G'-band) will give better conductivity of the SWCNT TCF. From this point of view, the Arc TCF is the best sample providing the highest conductivity in comparison to TCFs made by other types of SWCNTs considered in this work, as can be seen from Fig. 3.6. However, it may be conjectured that different optimization conditions for dispersion and film preparation may change the TCF properties. In spite of such a possibility, the argument for the material quality dependence described above still holds true. The trend of the change in sheet conductance is similar at different transmittance regions, although the slopes are different. This suggests that our definition of the effective material parameter is quite understandable from a material point of view to describe the TCF performance.

The film morphologies of the SWCNT network of TCFs are presented in Fig. 3.8. All other processing parameters except types of SWCNTs were kept constant through the TCFs preparation process. It is noted that the film morphologies of the CVD and the HiPCO TCFs were significantly different from those of the Laser and the Arc samples. The circular spots were visible in the CVD and the HiPCO samples, as can be seen in Fig. 3.8a–b. These are the remaining SDS surfactants in the film in spite of the repeated washing process. This resulted presumably from

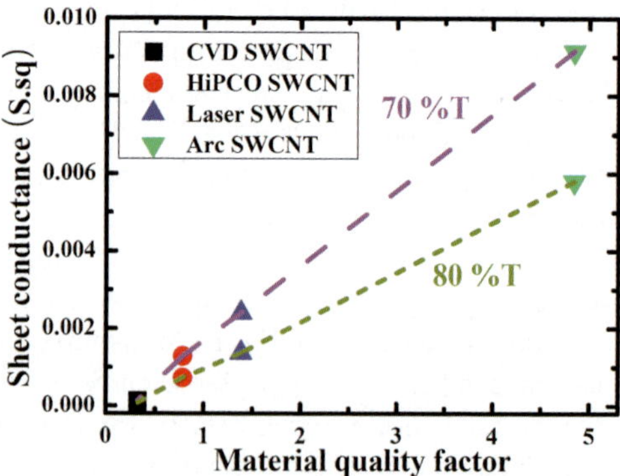

Fig. 3.7 The sheet conductance changes as a function of the material quality factor at different transmittance of 70 and 80%

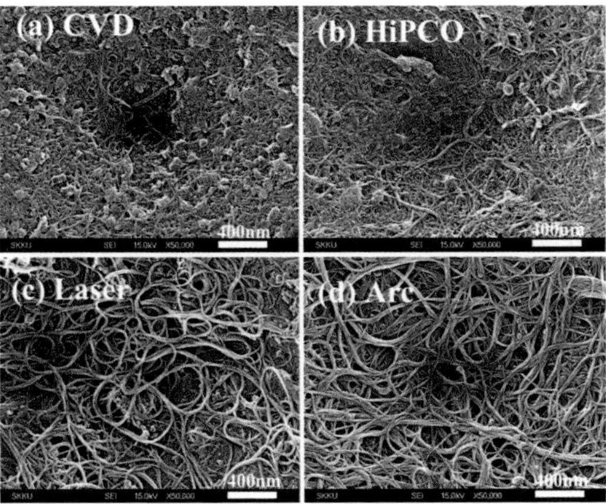

Fig. 3.8 FE-SEM images of the as-prepared TCFs using different types of SWCNTs (scale bar: 400 nm): (a) CVD SWCNTs, (b) HiPCO SWCNTs, (c) Laser SWCNTs, and (d) Arc SWCNTs

the stronger binding of surfactants with defects on the SWCNT walls that were hardly removable in the current process compared to the Laser and the Arc samples. Another important feature is the sparseness and networking capability of the CNT bundles. The Laser and the Arc films revealed low CNT density but better network formation by an entanglement of the SWCNT bundles. On the other hand, the CNT density in the CVD and the HiPCO films was high but with relatively less degree of entanglement. As a consequence, the transmittance in the CVD and the HiPCO films was significantly dropped at high density of the CVD and HiPCO films. From the SWCNT density and networking capability of the CNT films, the Arc sample again seemed to be the best material for TCFs among these samples we studied in this work.

3.4 Conclusion

In summary, the flexible TCFs with various types of SWCNTs have been investigated by using spray method. The characteristics of TCFs were analyzed by both the material quality factors and the film quality factors. We extracted all the material parameters of SWCNT powders such as diameter, purity and burning temperature, D-band and G'-band in Raman spectra. We found that G'-band represented well the metallicity of SWCNTs. We defined the material quality factor that dominated the conductivity of SWCNTs by using the diameter, purity, and the average of intensity ratio of G'-band to D-band. The sparseness of the SWCNT network and the degree of entanglement in the film were also important film parameters to determine the conductivity of the film. By taking account of all these associated effects, we were able to provide an explanation of the SWCNT film performance. A further improvement in the film performance could be possible by choosing highly conductive CNTs evaluated by material quality factor and by enhancing the degree

Sreekumar, T.V., Liu, T. and Kumar, S., 2003, Single-wall Carbon nanotube films. *Chemistry of Materials*, **15**, pp. 175–178.

Sun, C.-H., Yin, L.-C., Li, F., Lu, G.-Q. and Cheng, H.-M., 2005, Van der Waals interactions between two parallel infinitely long single-walled nanotubes. *Chemical Physics Letters*, **403**, pp. 343–346.

Thess, A., Lee, R., Nikolaev, P., Dai, H., Petit, P., Robert, J., Xu, C., Lee, Y. H., Kim, S. G., Rinzler, A. G., Colbert, D. T., Scuseria, G. E., Tománek, D., Fischer, J. E. and Smalley, R. E., 1996, Crystalline ropes of metallic carbon nanotubes. *Science*, **273**, pp. 483–487.

Unalan, H. E., Fanchini, G., Kanwal, A., Pasquier, A. D. and Chhowalla, M., 2006, Design criteria for transparent single-wall carbon nanotube thin-film transistors. *Nano Letters*, **6**, pp. 677–682.

Vaisman, L., Marom, G. and Wagner, H. D., 2006, Dispersions of surface-modified carbon nanotubes in water-soluble and water-insoluble polymers. *Advanced Functional Materials*, **16**, pp. 357–363.

Wu, Z. C., Chen, Z., Du, X., Logan, J. M., Sippel, J., Nikolou, M., Kamaras, K., Reynolds, J. R., Tanner, D. B., Hebard, A. F. and Rinzler, A. G., 2004, Transparent, conductive carbon nanotube films. *Science*, **305**, pp. 1273–1276.

Zhang, M., Fang, S., Zakhidov, A. A., Lee, S. B., Aliev, A. E., Williams, C. D., Atkinson, K. R. and Baughman, R. H., 2005, Strong, transparent, multifunctional, carbon nanotube sheets. *Science*, **309**, pp. 1215–1219.

Zhu, H., Zhao, G.-L., Masarapu, C., Young, D. P. and Wei, B., 2005, Super-small energy gaps of single-walled carbon nanotube strands. *Applied Physics Letters*, **86**, pp. 203107.

Chapter 4
Raman Spectroscopy on Double-Walled Carbon Nanotubes

Wencai Ren and Hui-Ming Cheng*

4.1 Introduction

Double-walled carbon nanotube (DWNT) is a special MWNT only consisting of two coaxial SWNTs. It is an ideal model to study the effect of interlayer interaction on the phonons and electronic structures of carbon nanotubes (CNTs), and have some unique properties and potential applications as nanodevices, due to its unique double wall structure. For example, it is expected to have some potential applications as molecular conductive wire or molecular capacitor in a memory device, depending on the electronic properties of the two constituent tubes (Saito et al., 1993). Recent studies also indicated that DWNTs are better field effect transistor (FET) channels than SWNTs (Shimada et al., 2004). Therefore, it is very important to realize the controllable synthesis of DWNTs and to identify the diameters and the electronic properties of the two constituent tubes for their promising applications as nano-scale electronic devices.

It is well established that every possible nanotube has a distinct electronic and vibrational spectrum, so that there is a one-to-one relation between the nanotube and the singularities in the 1D joint electronic density of states (JDOS) (Dresselhaus et al., 2002). Moreover, the electronic states are highly sensitive to the diameter of nanotubes. In the resonant Raman effect, a large enhancement in the Raman signal occurs only when the incident or scattered photon is in resonance with a singularity in the 1D JDOS of nanotubes (Dresselhaus et al., 2002). Therefore, resonance Raman spectroscopy can be used as a sensitive probe of the geometrical and electronic structure of CNTs through coupling between electrons and phonons in this one-dimensional system.

In this review, we will present our recent progress on the controllable synthesis and detailed structural characterization of DWNTs. A floating catalyst chemical vapor deposition (CVD) method was proposed for the synthesis of DWNTs

---

*Shenyang National Laboratory for Materials Science, Institute of Metal Research, Chinese Academy of Sciences, Shenyang 110016, China
e-mail: cheng@imr.ac.cn

Z. Tang, P. Sheng (eds), *Nano Science and Technology.*
© Springer 2008

with high purity, good alignment and narrow diameter distribution with a relatively large scale. Moreover, we performed systematical Raman studies on the as-prepared DWNTs to probe their detailed structure characteristics and several unique Raman features were found.

4.2 Experimental

The synthesis equipment was described in detail in our previous paper (Cheng et al., 1998a). In general, methane was used as the carbon source, hydrogen as the carrier gas, ferrocene as the catalyst precursor, and thiophene was carried into the reactor by methane to enhance the growth of CNTs. For the DWNT growth, the reaction tube was first heated to reach 1373 K in a hydrogen atmosphere. Then, methane and hydrogen, accompanying with the catalyst precursor and thiophene, were introduced into the reaction tube for 10 min, followed by cooling the CVD system in hydrogen to room temperature. Depending on the experimental conditions, web-like DWNTs (Ren et al., 2002) or macroscopically long DWNT strands (Ren and Cheng, 2005) as long as 10 cm can be selectively formed during the reaction process. It is worthy of note that this method can be used to synthesize DWNTs in a large scale due to the continuous introduction of the carbon source, carrier gas, catalysts and thiophene.

The morphology and microstructure of the product were observed using scanning electron microscopy (SEM), high-resolution transmission electron microscopy (HRTEM JOEL 2010 200kV) and Raman spectroscopy (Jobin Yvon LabRam HR800, excited by 632.8 nm laser). It should be pointed out that all Raman spectra were taken in backscattering configuration, with the incident and scattered light propagating perpendicular to the rope axis (Ren et al., 2002, 2005, 2006). For the polarized Raman studies on the G band of DWNT ropes, the spectra were recorded with parallel VV and crossed VH polarizations of the incoming and scattered light. For the related discussions on the selection rules, we define Z as the DWNT axis direction and Y as the photon propagation direction.

4.3 Results and Discussion

4.3.1 Synthesis of DWNTs

Figure 4.1 and Fig. 4.2 show, respectively, SEM, TEM and HRTEM images of the as-prepared web-like DWNTs (Ren et al., 2002) and DWNT ropes (Ren and Cheng, 2005). It is apparent that the products are very pure with less amorphous carbon and catalyst particles attached, which is a common characteristic for the nanotubes produced by the floating catalyst CVD method. In the DWNT ropes, DWNTs grew along one direction, which was parallel to the direction of the flow gas in the reaction system. Therefore, we infer that the orientation force for the growth of the

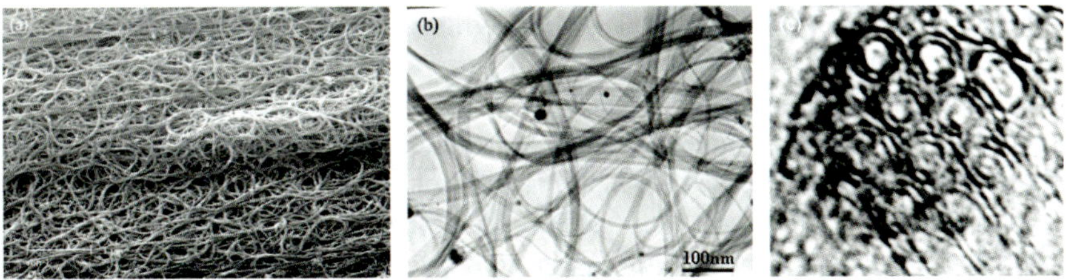

Fig. 4.1 SEM, TEM and HRTEM images of the web-like DWNTs (Ren et al., 2002)

long DWNT ropes is from the gas flow. Another point we need point out is that the diameter distribution of DWNTs prepared by our method is much narrower than those prepared by hydrogen arc discharge method (Hutchison et al., 2001). For the web-like product, the inner and outer diameters of DWNTs are in the range of 1.0–2.0 nm and 1.6–2.8 nm, respectively. For the DWNT ropes, the inner and outer diameter of DWNTs are dominantly in the range of 1.0–1.3 nm and 1.7–2.0 nm, respectively, which is approach to those prepared by C_{60} coalescence in SWNTs at high temperature (Bandow et al., 2001). The successful synthesis of these nanotubes opens up the possibility for their further property studies and applications.

4.3.2 Radial Breathing Mode of DWNTs

It is well established that the radial breathing mode (RBM) is a particularly valuable tool for detecting SWNTs and measuring the diameter of SWNTs (Dresselhaus et al., 2002). Therefore, probing and analyzing the RBM feature of DWNTs is a key for their further structural characterization. We consider that the effects of coupling between adjacent graphene layers of DWNTs on the vibrational and electronic states are not of sufficient strength to significantly affect the physical properties of each constituent due to the weak van der Waals interaction, as similar to the intra-bundle and inter-tube coupling in a SWNT bundle (Dresselhaus and Eklund, 2000). Therefore, it is not expected that many new Raman modes actually be observed, and the spectral intensity should still be dominated by the selection rules for the two constituent SWNTs.

The Raman spectra taken from the web-like sample are shown in Fig. 4.3. It is worth noting that different from that observed for SWNTs, the RBM band of

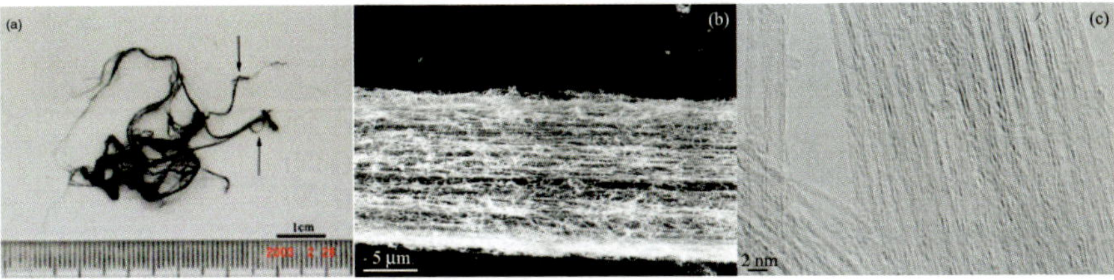

Fig. 4.2 Optical photo, and SEM, HRTEM images of DWNT ropes (Ren and Cheng, 2005)

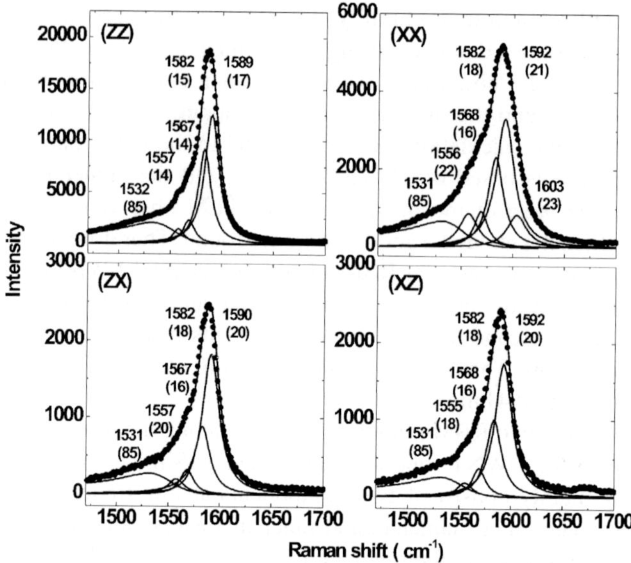

Fig. 4.5 Lineshape analysis for the four different polarized spectra of the DWNT bundles. The frequencies (widths) of the modes in the figure are displayed in cm^{-1} (Ren et al., 2005)

As stated above, for DWNTs, it is assumed that the interlayer interactions are similar to the intertube interactions in a SWNT bundle, so that the free nanotube symmetries can also be used to analyze the G-band feature of DWNTs (Dresselhaus et al., 2002; Jorio et al., 2000). Figure 4.5 shows the lineshape analysis results on the G band of DWNTs in different polarized geometries. It is noteworthy that all the modes exhibit different polarization-dependent relative intensities (see Fig. 4.5 and Table 4.2), suggesting that they are all intrinsic to the DWNTs.

The analysis on the RBM band of aligned DWNT ropes indicates that the inner and outer tubes excited by 632.8 nm laser are metallic and semiconducting, respectively. This result provides an useful information on the symmetry assignment of the G band of DWNTs. Based on symmetry analysis on SWNTs, the low frequency mode $1531 \, cm^{-1}$ was firstly assigned to the A (A_{1g}) mode of the metallic inner

Table 4.2 Relative intensities of the G-band modes belonging to different irreducible representations (IR) in the different polarized Raman spectra of DWNTs. For the modes at 1567, $1591 \, cm^{-1}$ and 1531, $1582 \, cm^{-1}$, we used their measured intensities in the (ZZ) polarized spectra as a reference, while for the modes at 1556 and $1603 \, cm^{-1}$, we used the (ZZ) intensity of the $1591 \, cm^{-1}$ mode as a reference (Ren et al., 2005)

$\omega_G \, (cm^{-1})$	IR	(ZZ)	(XX)	(ZX)	(XZ)
1556	E_2	0.12	0.07	0.02	0.02
1567	$A_1 + E_1$	1.00	0.40	0.14	0.16
1591	$A_1 + E_1$	1.00	0.27	0.15	0.14
1603	E_2	0.00	0.07	0.00	0.00
1531	A_1	1.00	0.32	0.13	0.12
1582	A_1	1.00	0.26	0.10	0.11

tubes because of the asymmetric BWF line shape. Generally, the ω_G^+ is $1590\,\mathrm{cm}^{-1}$ for S-SWNT bundles, and $1582\,\mathrm{cm}^{-1}$ for M-SWNT bundles, which is independent of d_t (Jorio et al., 2000; Brown et al., 2001). Therefore, the mode $1582\,\mathrm{cm}^{-1}$ with a Lorentzian line shape was assigned to high frequency A (A_{1g}) mode of the metallic inner tubes. Moreover, the intensity ratios $I_{(ZZ)}{:}I_{(XX)}$ for the above two modes are consistent with the theoretical value of 1.00:0.25. The residual modes with Lorentzian line shape were all assigned to originate from the semiconducting outer tubes. Our experimental results show that the highest frequency $1603\,\mathrm{cm}^{-1}$ tangential mode only appears with a relatively low intensity in the (XX) parallel-polarized spectra (see Fig. 4.5 and Table 4.2). This result is in agreement with group theory predictions for the E_2 (E_{2g}) mode. The $1556\,\mathrm{cm}^{-1}$ mode was also assigned to E_2 (E_{2g}) symmetry of the semiconducting outer tubes, because of its lower intensity in all the polarization configurations and the approximately equal frequency shifts with $1603\,\mathrm{cm}^{-1}$ from the central frequency of $1580\,\mathrm{cm}^{-1}$ for graphite. The peaks at 1567 and $1591\,\mathrm{cm}^{-1}$ exhibit a comparable intensity ratio $I_{(ZZ)}{:}I_{(XX)}$ to the theoretical value of 1.00:0.25 for A symmetry modes (Saito et al., 1998). Moreover, the cross-polarized spectra also exhibit the above two features. Therefore, we assigned them to two A (A_{1g}) $+$ E_1 (E_{1g}) modes of the semiconducting outer tubes.

To further determine the effect of interlayer interaction on the G band, we performed a comparison of mode symmetry and frequency for SWNTs and DWNTs with a similar diameter and the same electronic property. Previous research results indicated that, for S-SWNT bundles with an average diameter of 1.85 nm, four peaks 1549, 1567, 1590, and $1607\,\mathrm{cm}^{-1}$ were found and respectively assigned to the E_2 (E_{2g}), two A (A_{1g}) $+$ E_1 (E_{1g}), and high frequency E_2 (E_{2g}) mode (Jorio et al., 2000). Our assignment results for the semiconducting outer tubes are consistent with the above results except for small frequency shifts due to the different average diameters. Moreover, it is worth noting that the parameters of the BWF lineshape for metallic inner tubes, such as the frequency and linewidth, are in good agreement with those for the M-SWNTs with similar diameters (Brown et al., 2001). All these results indicate that the two constituent tubes independently retain their original mode symmetries and vibrations, consistent with the theoretical predictions (Damnjanovic et al., 2003) and above assumption.

4.3.4 D Band and G′ Band of DWNTs

For a SWNT, it was well established that its structure can be uniquely determined by its (n, m) indices (Dresselhaus et al., 2002). However, a DWNT cannot be determined only by two different (n, m) combinations, but also by the atomic arrangement of its two constituent tubes, i.e., their relative circumferential rotation angle and their axial translation is an important determining parameter (Saito et al., 1993). Moreover, theoretical calculations for energy bands of DWNTs have concluded that their electronic structures are sensitive to their structural symmetry, that is, the atomic correlation between adjacent nanotubes (Saito et al., 1993, and Sanvito et al., 2000).

anticrossings of energy bands due to the interlayer interaction (Saito et al., 1993). The occurrence of additional vHSs makes it possible to give rise to some additional vHSs satisfying the triple resonance conditions. Moreover, the very singular density of states makes related D-band frequencies distinguishable due to the resulting very narrow linewidth. Consequently, some new peaks were expected to appear in the D band of commensurate DWNTs, having different frequencies with those of incommensurate DWNTs. Therefore, the two observed outer peaks in the D band can be deduced to originate from the occurrence of additional vHSs related to a stronger interlayer interaction for the commensurate DWNTs in the sample. The above analysis can be further proved by the recent studies on the D band and G' band of graphene, where only one peak is found for the monolayer graphene, and four peaks are observed for the bi-layer graphene with AB stacking due to the modification of electronic structures by strong interlayer interaction (Ferrari et al., 2006).

4.4 Conclusion

A brief review about the synthesis and Raman spectroscopy of DWNTs is here presented, emphasizing in particular how such spectra can be observed for DWNTs and the use of these spectra for the further structural characterization of DWNTs. A floating CVD method was proposed for the synthesis of DWNTs with high purity, good alignment and a very narrow diameter distribution. Double-peak feature was found in the RBM of DWNTs, which is considered to be associated with the outer and inner diameter, respectively. From the polarized Raman studies on the G band of DWNTs, it was found that the two constituent tubes of a DWNT independently remain their symmetries and vibrations. Four well-resolved peaks with very narrow linewidths were found in the D-band and G'-band features of double-walled carbon nanotubes (DWNTs). Combined with theoretical predictions, we constitute the first Raman evidence for atomic correlation and the resulting electronic structure modification of the two constituent tubes in DWNTs. However, we must point out that the obtained results about Raman spectroscopy of DWNTs are based on bundles and further investigations on individual nanotubes are required to justify these deductions.

Acknowledgments This work was supported by National Natural Science Foundation of China (No. 90606008), Ministry of Science and Technology (No. KJCX2-YW-M01), and Chinese Academy of Sciences 2006CB932701.

References

Bandow, S., Takizawa, M., Hirahara, K., Yudasaka, M., and Iijima, S., 2001, Raman scattering study of double-wall carbon nanotubes derived from the chains of fullerenes in single-wall carbon nanotubes. *Chemical Physics Letters*, **337**, pp. 48–54.

Brown, S.D.M., Jorio, A., Corio, P., Dresselhaus, M.S., Dresselhaus, G., Saito, R., and Kneipp, K., 2001, Origin of the Breit-Wigner-Fano lineshape of the tangential G-band feature of metallic carbon nanotubes. *Physical Review B*, **63**, p. 155414.

Cheng, H.M., Li, F., Su G., Pan, H.Y., He, L.L., Sun, X., and Dresselhaus, M.S., 1998a, Large-scale and low-cost synthesis of single-walled carbon nanotubes by the catalytic pyrolysis of hydrocarbons. *Applied Physics Letters*, **72**, pp. 3282–3284.

Cheng, H.M., Li, F., Sun, X., Brown, S.D.M., Pimenta, M.A., Marucci, A., Dresselhaus, G., and Dresselhaus, M.S., 1998b, Bulk morphology and diameter distribution of single-walled carbon nanotubes synthesized by catalytic decomposition of hydrocarbons. *Chemical Physics Letters*, **289**, pp. 602–610.

Damnjanovic, M., Dobardzic, E., Milosevic, I., Vukovic, T., and Nikolic, B., 2003, Lattice dynamics and symmetry of double wall carbon nanotubes. *New Journal of Physics*, **5**, pp. 148.

Dresselhaus, M.S. and Eklund, P.C., 2000, Phonons in carbon nanotubes. *Advances in Physics*, **49**, pp. 705–814.

Dresselhaus, M.S., Dresselhaus, G., Jorio, A., Souza Filho, A.G., and Saito, R., 2002, Raman spectroscopy on isolated single wall carbon nanotubes. *Carbon*, **40**, pp. 2043–2061.

Ferrari, A.C., Meyer, J.C., Scardaci, V., Casiraghi, C., Lazzeri, M., Mauri, F., Piscanec, S., Jiang, D., Novoselov, K.S., Roth, S., and Geim, A.K., 2006, Raman spectrum of graphene and graphene layers. *Physical Riview Letters*, **97**, pp. 187401.

Hutchison, J.L., Kiselev, N.A., Krinichnaya, E.P., Krestinin, A.V., Loutfy, R.O., Morawsky, A.P., Muradyan, V.E., Obraztsova, E.D., Sloan, J., Terekhov, S.V., and Zakharov, D.N., 2001, Double-walled carbon nanotubes fabricated by a hydrogen arc discharge method. *Carbon*, **39**, pp. 761–770.

Jorio, A., Dresselhaus, G., Dresselhaus, M.S., Souza, M., Dantas, M.S.S., Pimenta, M.A., Rao, A.M., Saito, R., Liu, C., and Cheng, H.M., 2000, Polarized Raman study of single-wall semiconducting carbon nanotubes. *Physical, Review Letters*, **85**, pp. 2617–2620.

Jorio, A., Fantini, C., Dantas, M.S.S., Pimenta, M.A., Souza Filho, A.G., Samsonidze, Ge, G., Brar, V.W., Dresselhaus, G., Dresselhaus, M.S., Swan, A.K., Ünlü, M.S., Goldberg, B.B., and Saito, R., 2002, Linewidth of the Raman features of individual single-wall carbon nanotubes. *Physical Review B*, **66**, p. 115411.

Ren, W.C. and Cheng, H.M., 2005, Aligned double-walled carbon nanotube long ropes with a narrow diameter distribution. Journal of Physical Chemistry B, **109**, pp. 7169–7173.

Ren, W.C., Li, F., and Cheng, H.M., 2005, Polarized Raman analysis of aligned double-walled carbon nanotubes. *Physical Review B*, **71**, p. 115428.

Ren, W.C., Li, F., Chen, J., Bai, S., and Cheng, H.M., 2002, Morphology, diameter distribution and Raman scattering measurements of double-walled carbon nanotubes synthesized by catalytic decomposition of methane. *Chemical Physics Letters*, **359**, pp. 196–202.

Ren, W.C., Li, F., Tan, P.H., and Cheng, H.M., 2006, Raman evidence for atomic correlation between the two constituent tubes in double-walled carbon nanotubes. *Physical Review B*, **73**, p. 115430.

Saito, R., Dresselhaus, G., and Dresselhaus, M.S., 1993, Electronic-structure of double-layer graphene tubules. *Journal of Applied Physics*, **73**, pp. 494–500.

Saito, R., Takeya, T., Kimura, T., Dresselhaus, G., and Dresselhaus, M.S., 1998, Raman intensity of single-wall carbon nanotubes. *Physical Review B*, **57**, pp. 4145–4153.

Sanvito, S., Kwon, Y.K., Tomanek, D., and Lambert, C.J., 2000, Fractional quantum conductance in carbon nanotubes. *Physical Review Letters*, **84**, pp. 1974–1977.

Shimada, T., Sugai, T., Ohno, Y., Kishimoto, S., Mizutani, T., Yoshida, H., Okazaki, T., and Shinohara, H., 2004, Double-wall carbon nanotube field-effect transistors: Ambipolar transport characteristics. *Applied Physics Letters*, **84**, pp. 2412–1414.

Souza Filho, A.G., Jorio, A., Dresselhaus, G., Dresselhaus, M.S., Saito, R., Swan, A.K., Ünlü, M.S., Goldberg, B.B., Hafner, H., Lieber, M., and Pimenta, A., 2001, Effect of quantized electronic states on the dispersive Raman features in individual single-wall carbon nanotube, *Physical Review B*, **65**, p. 035404.

Chapter 5
Interface Design of Carbon Nano-Materials for Energy Storage

Feng Li*, Hong-Li Zhang, Chang Liu and Hui-Ming Cheng

5.1 Introduction

Lithium ion batteries (LIB) have been a forerunner and market leader since late 1990s in the field of rechargeable battery industry especially for portable electronic devices, and are also very attractive in electric vehicles and hybrid electric vehicles. In order to achieve high-performance LIB, developing novel electrode materials, on one hand, is pivotal; on the other hand, a deeper understanding of the related electrochemical phenomena along with electrode reactions is definitely necessary (Tarascon and Armand, 2001).

It is well known that the interface between electrode materials and electrolytes plays an important role in affecting the first columbic efficiency, cyclability, rate capability, and safety of LIB (Aurbach, 2003). Especially for carbonaceous anode materials widely used in current LIB, there is a kind of solid electrolyte interphase (SEI) film formed on the surface due to electrolytes decomposition (Balbuena and Wang, 2004). The SEI film is insulating for electrons but conductive for Li ions, and can significantly influence the cyclability of carbon anode. Here, we introduce the formation and evolution of SEI film on the surface of natural graphite (NG) spheres, which show obvious advantages over artificially graphitized meso-phase carbon micro-beads owing to low cost and intrinsically high crystallinity (Vetter et al., 2005). Based on the interface results, we propose core-shell design to modify the electrode materials for LIB by the chemical vapour deposition process. We can obtain the full coating amorphous carbon core-shell structure without the catalyst and nanotube/nanofiber coating nano/micro urchin-like structure with catalyst by chemical vapor deposition (Zhang et al., 2005, 2006a,b).

5.2 Experimental

An electrochemical cell was assembled in Ar-filled glove box (Mbraun, Unilab, H_2O and $O_2 <1ppm$) with a working electrode composed of anode materials (85 wt%),

---

* Shenyang National Laboratory for Materials Science, Institute of Metal Research, Chinese Academy of Sciences, Shenyang 110016, China
e-mail: fli@imr.ac.cn

carbon black (5 wt%) and poly(vinylidene fluoride) (PVDF) binder (10 wt%), a lithium foil counter electrode and a porous separator (Celgard, 2400). The electrolyte was 1 M $LiPF_6$ in a mixture of EC/DMC (1:1 volume ratio) (battery grade). The cells were then galvanostatically charged and discharged for certain cycles in the voltage range of 0.001–2.5 V versus Li/Li^+ at a current density of $0.2\,mA/cm^2$.

The detailed description on the characterization of cycled electrodes by focused ion beam (FIB) workstation and the preparation of the core-shell composite anode materials can be found in (Zhang et al., 2005, 2006a,b).

5.3 Results and Discussion

5.3.1 SEI Film Observations

Figure 5.1a shows the SEM image of a typical original natural graphite sphere, we can see that there are many cracks and flakes on the surface. Figure 5.1b–d present the secondary electron FIB images and corresponding elemental line scan analysis (ELSA) curves of a NG sphere that was from a working electrode discharged to 0 V from OCV. In Fig. 5.1b, a rough film can be obviously discerned on the surface of the NG sphere, which shields the original surface cracks as shown in Fig. 5.1a. Moreover, a small split (Fig. 5.1c) on the film is also observed. The split is possibly due to the increase of internal pressure from gas production causing the expansion of the graphite sphere. The ELSA curves in Fig. 5.1d indicate the relative content variation of fluorine (F), oxygen (O), carbon (C) and phosphorus (P) along the black line in Fig. 5.1c. According to the variation of peaks and valleys in the ELSA curves, it can be concluded that the surface species of the freshly formed SEI film are distributed non-uniformly, which is due to different reactivity toward electrolytes on different local parts of NG spheres.

With the aid of FIB workstation, we successfully cut the NG sphere as shown in Fig. 5.1b, and obtain a clear cross-section image as displayed in Fig. 5.2a. The thickness of SEI film on the surface is directly measured, and an element analysis is performed inside the sphere. The results of ELSA curves demonstrate that there are peaks of F and O around the internal cracks, which reveals that electrolytes can penetrate into the inside of NG spheres through the surface cracks and decompose there to form a kind of internal SEI film.

Figure 5.3 presents the FIB image and ELSA curves of a NG sphere experienced 24 discharge/charge cycles. It can be seen that the SEI film splits upon repeated cycles due to the increase of internal pressure from gas (e.g. C_2H_2) production inside the sphere and of volume change stress caused by intercalation/deintercalation of Li^+. The ELSA curves (Fig. 5.3b) show alternate peaks and valleys around the cracks, and the average content of each element keeps nearly horizontal in the whole scanning range. The result indicates that the SEI film becomes relatively uniform in composition after reaching its stable state.

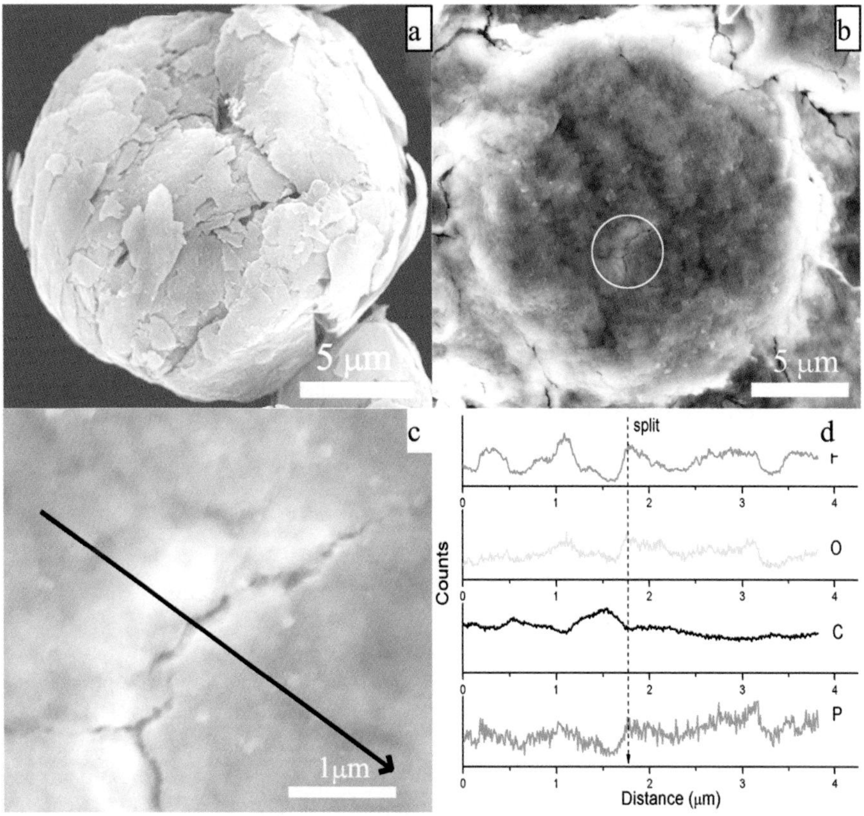

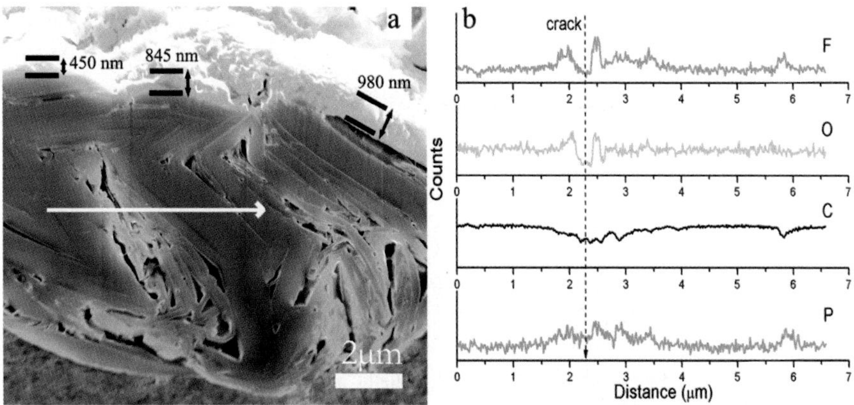

Fig. 5.1 (a) SEM image of a typical NG sphere; (b and c) FIB secondary electron images showing the surface of a NG sphere which have experienced the first discharge process; (d) ELSA curves along the black line in (c) (Zhang et al., 2005)

Fig. 5.2 (a) A secondary electron FIB image showing the cross section of a NG sphere experienced the first discharge cycle; (b) ELSA curves along the white line in (a) (Zhang et al., 2005)

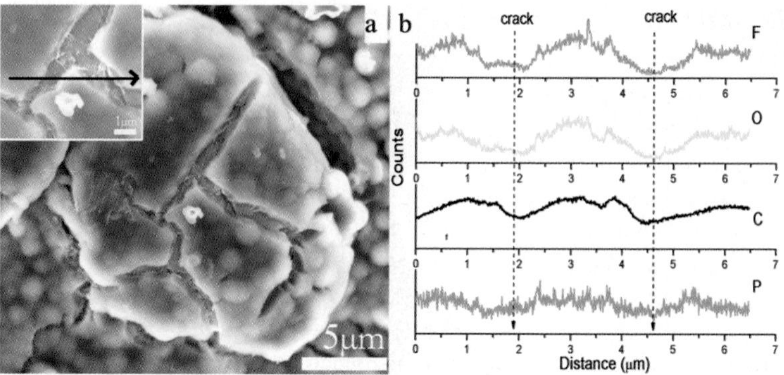

Fig. 5.3 (a) A secondary electron FIB image showing the changes of surface SEI film of a NG sphere experienced 24 cycles; (b) ELSA curves along the white line in (a) (Zhang et al., 2005)

5.3.2 Electrode Materials Design

In order to avoid the occurrence of internal SEI film and the formation of a thick SEI film, the core-shell structure can be used as controlling the SEI film. Therefore, we propose core-shell design to modify the electrode materials for LIB by the chemical vapour deposition process (Fig. 5.4). The advantages of this design are (i) forming a continuous conductive network in the bulk of electrodes to decrease electrode polarization; (ii) improving the adsorption and immersion of electrolytes on the surface of electro- active materials to facilitate the electrode reaction kinetics; (iii) as a buffer among electroactive materials due to their superior resiliency; (iv) being a kind of electroactive material itself without lowering the whole capacity of anode materials notably.

We obtained pyrolytic carbon coated NG core-shell structure and nanotube/nanofiber coating nano/micro urchin-like structure with catalyst by chemical vapor deposition.

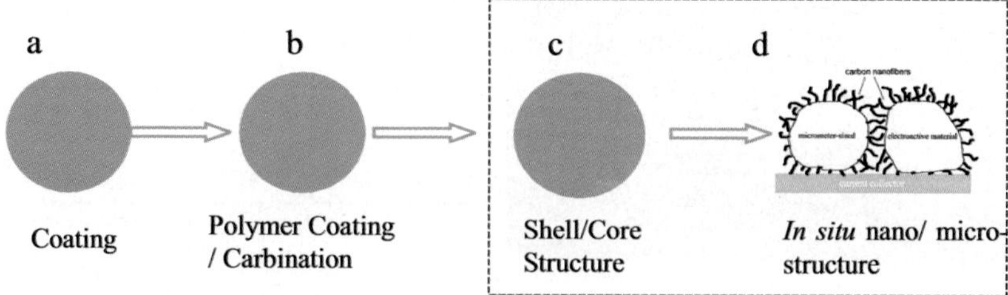

Fig. 5.4 The evolution of electrode materials of LIB

5.3.3 Characterizations of Electrode Materials

Figure 5.5a and b presents more direct evidence about the existence of pyrolytic carbon layer by the cross-section images of a coated NG sphere, which was obtained using a FIB workstation that has the ability to accurately cut a specimen (in the scale of micrometers or nanometers) at any user-defined position and clearly display corresponding secondary electron images. An obvious layer in white contrast can be identified at the periphery of the NG sphere, indicating that the pyrolytic carbon layer has a thickness of 250 nm. According to the thickness and some other basic data such as the average diameter of NG spheres, the densities of NG spheres and pyrolytic carbon, one can estimate that the approximate weight percent of deposited carbon was 13%, which was very close to the result (15%) obtained by a thermogravimetric analysis (TG) method. Meanwhile, many cracks inside the graphite sphere are also clearly shown in Fig. 5.5a, which are unavoidable during the sphere-making process from common flaky graphite. NG spheres were coated in one step by a uniform layer of pyrolytic carbon with a thickness of 250 nm. The coated sample as anode material of a lithium-ion battery displayed excellent cyclability, withthe capacity above 320 mAh/g at the 15th cycle in comparison with less than 200 mAh/g for the original NG spheres. Meanwhile, the first CE was also improved to 88% from original 80%. The remarkably improved electrochemical properties for the coated NG spheres can be mainly attributed to the following factors: the lowered specific surface area, the inhibition of internal SEI film formed around the internal cracks, and the formation of a thin and compact SEI film on the outer surface of the coated NG spheres.

In order to avoid the occurrence of internal SEI film and the formation of a thick SEI film, we propose to grow CNFs on the surface of NG spheres in situ by catalytic chemical vapor deposition. In this way, a nano/micro hybrid urchin-like composite (NG/CNF) is achieved as shown in Fig. 5.6a, where NG spheres act as inner cores and CNF as surrounding spines. From the Raman spectra in Fig. 5.6b, it can be noted that the grown-CNF is of non-graphitic. Therefore, the urchin-like composite is also a combination of graphitic (core) and non-graphitic (spines) structure. Figure 5.6c

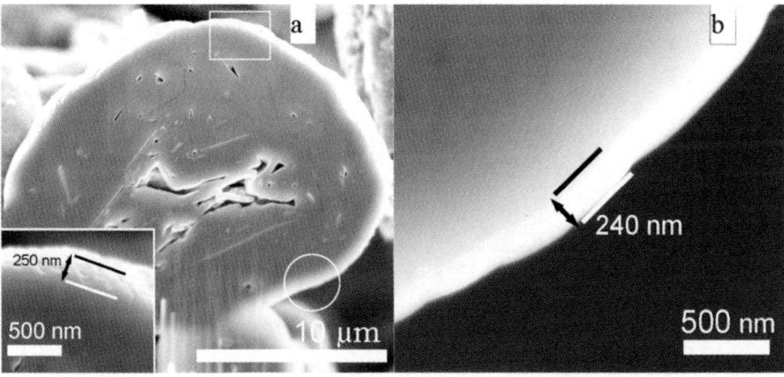

Fig. 5.5 Secondary electron FIB images showing the cross section of a coated NG sphere. The inset in panel (a) is the magnified image of the rectangle; panel (b) is the magnified image of the circle area in panel (a). (Zhang et al., 2006a).

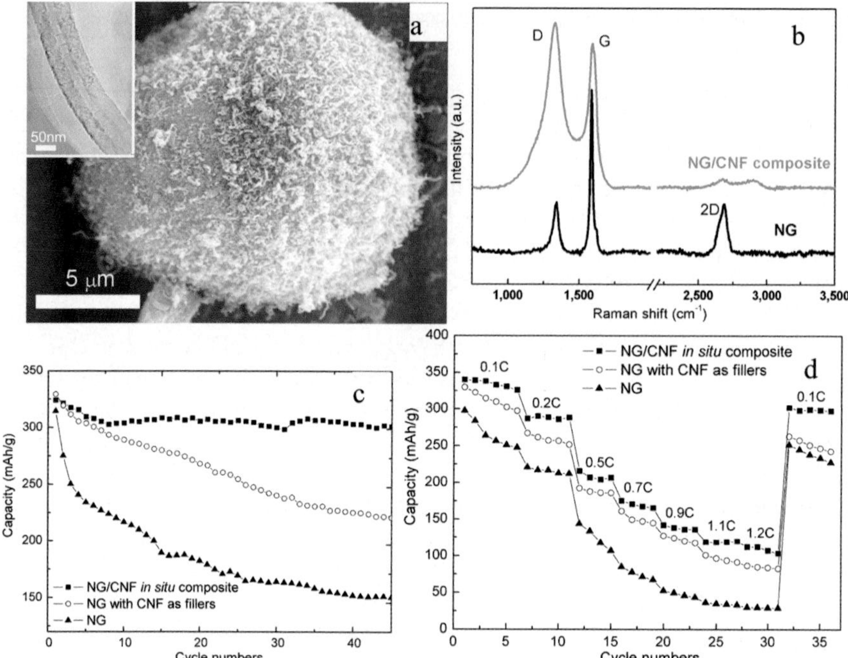

Fig. 5.6 (a) A SEM image showing the urchin-like NG/CNF in situ composite; (b) Raman spectra of the pristine NG and the NG/CNF in situ composite; (c and d) Comparison of cyclability and rate capability among the pristine NG, the NG with CNF as direct fillers, and the NG/CNG in situ composite (Zhang et al., 2006b)

displays the comparison of cyclability among the pristine NG spheres, the NG with CNF as direct fillers, and the NG/CNF in situ composite. It is clearly seen that a great improvement in cycling performance is achieved for the in situ composite, the capacity of which is still kept at above 300 mAh/g up to 45 cycles, about 150 mAh/g higher than that of the pristine NG. On the contrary, the capacity of the pristine NG and the NG with CNF as fillers drops continuously in the whole cycles. Figure 5.4b displays rate capability at different C-rates ranging from 0.1 to 1.2 C. Again, the NG/CNF in situ composite shows a better behavior than the pristine NG and the NG with CNF as fillers.

5.4 Conclusion

The interfaces with electrolytes are key factors for carbon nano-materials in their applications of energy storage. In this report, the formation and evolution of the SEI film on the surface of natural graphite spheres in the electrolyte of 1 M $LiPF_6$ in ethylene carbonate and diethyl carbonate were investigated by using FIB technique. A concept of "internal SEI film" is first proposed based on the characterization of the cross section of the natural graphite spheres with the aid of FIB. Secondly, based on the results from interface investigation, urchin-like nano/micro hybrid materials are designed and achieved to modify conventional micrometer-sized electroactive materials for LIB. Moreover, a pyrolytic carbon coated NG core-shell structure and a nanotube/nanofiber coated NG nano/micro urchin-like structure were prepared. These composite anode materials display significantly improved cyclability and rate capability in comparison with the pristine NG. We can obtain the full coating amorphous carbon core-shell structure without the catalyst and nanotube/nanofiber coating nano/micro urchin-like structure with catalyst by chemical vapor deposition. Electrochemical measurements indicate that the cyclability and rate capability of the composite as anode material for LIB are significantly improved. Furthermore, the design also demonstrates its effectiveness in other kinds of anode and cathode materials such as transition metal oxides. These results may be inspirable for exploring high performance electrode materials.

Acknowledgments This work was supported by National Natural Science Foundation of China (No. 50632040, 50328204).

References

Aurbach, D., 2003, Electrode–solution interactions in Li-ion batteries: a short summary and new insights. *Journal of Power Sources*, 119–121, pp. 497–503.

Balbuena, P.B., Wang, Y.X. (eds) 2004, *Lithium-Ion Batteries Solid-Electrolyte Interphase* (Imperial College Press).

Tarascon, J.-M. and Armand, M., 2001, Issues and challenges facing rechargeable lithium batteries. *Nature*, **414**, p. 359

Vetter, J., Novák, P., Wagner, M.R., Veit, C., Möller, K.-C., Besenhard, J.O., Winter, M., Wohlfahrt-Mehrens, M., Vogler, C. and Hammouche, A., 2005, Ageing mechanisms in lithium-ion batteries. *Journal of. Power Sources, 147*, p. 269

Zhang, H. L., Li, F., Liu, C., Tan, J., Cheng, H. M., 2005, New insight into the solid electrolyte interphase with use of a focused ion beam. *Journal of Physical Chemistry B*, **109**, pp. 22205–22211.

Zhang, H.L., Liu, S.H., Li, F., Bai, S., Liu, C., Tan, J., Cheng, H.M., 2006a, Electrochemical performance of pyrolytic carbon-coated natural graphite spheres. *Carbon*, **44**, pp. 2212–2218.

Zhang, H.L., Zhang, Y., Zhang, X.G., Li, F., Liu, C., Tan, J., Cheng, H.M., 2006b, Urchin-like nano/micro hybrid anode materials for lithium ion battery, *Carbon*, **44**, pp. 2778–2784.

SWNTs have higher quality and density (Li et al., 2004b; Zhai et al., 2006a,b). These works led us to explore the carbonization mechanism of organic species in CoAPO-5 crystal channels in more detail and optimize SWNTs synthesis process further.

In this paper, we report the occurred forms and detailed decomposition process of organic molecules TPA confined in the channels of CoAPO-5 crystals. By coupling micro-Raman spectrometry and mass spectrometry, it has been possible to show that tripropylammonium cation decomposition occurs via abstraction of propylene and stepwise formation of dipropyl- and n-propylammonium cations. Micro-Raman measurements at various temperatures show that the 0.4 nm nanotubes are formed at about 673 K. Compared with that of $AlPO_4$-5 single crystal, the carbonization temperature of organic species inside the channels of CoAPO-5 crystals is notably decreased.

6.2 Experimental

An AFI crystal is a type of microporous aluminophosphate crystallites. Its framework consists of alternating tetrahedral $(AlO_4)^-$ and $(PO_4)^+$, which form parallel open channels packed in the hexagonal structure. The inner diameter of the 12-ring channel is 0.73 nm, and the distance between two neighboring parallel channels is 1.37 nm. During the hydrothermal growth of the AFI crystals, TPA molecules were introduced into the channels as growth template. The TPA molecules are oriented in a head-tail manner along the [001] direction of the channels (Qiu et al., 1989). In this work, CoAPO-5 single crystals were synthesized by the hydrothermal method. Cobalt acetate, aluminum tri-isopropoxide $[(iPrO)_3Al]$ and phosphoric acid (H_3PO_4 85 wt. %) were used as cobalt, aluminum and phosphorus sources, respectively. The aluminum tri-isopropoxide and cobalt acetate were first hydrolyzed in water, and then H_3PO_4, and TPA template was dropwisely added into the aluminum tri-isopropoxide solution under vigorous stirring. The gel formed from the final reaction mixture was sealed in a Teflon-lined stainless autoclave and heated at 448 K for 24 hours. The solid products were filtered, washed with distilled water, and dried at 353 K in atmosphere.

The composition of the CoAPO-5 crystal was analyzed by X-ray fluorescence (XRF) using element analyzer JSX-3201Z. The volatiles evolved from CoAPO-5 crystals during pyrolysis process were measured by an ABB mass spectrometer. In order to eliminate the influence of physisorbed water inside the channels, the as-synthesized sample is firstly treated under a vacuum of 10^{-3} mbar at 373 K for 3 hours. After the water was desorbed, these crystals were treated at temperatures between 373 and 873 K, with a heating rate of 3 K/min. Raman spectra of the samples were measured at various temperatures by Jobin Yvon-T64000 micro-Raman spectroscopy using 514.5 nm line of an Ar ion laser excitation. The equipped CCD detector was cooled using liquid nitrogen. The incident laser light was polarized parallel to the tube axis.

6.3 Results and Discussion

The as-grown CoAPO-5 crystals have typical size of $\sim 50\,\mu m$ in diameter and $\sim 400\,\mu m$ in length. As confirmed by X-ray diffraction and FTIR, the CoAPO-5 crystal has the same crystal structure as that of $AlPO_4$-5 crystals. The composition of crystal is tested by X-ray fluorescence (XRF). Figure 6.1 shows the composition of the CoAPO-5 crystal plotted as a function of the cobalt molar ratio in the starting gel. The vertical axes are the stoichiometric ratio of Al, P (left) and Co (right) in the CoAPO-5 crystals. The analytical data indicated an increase in the cobalt content in the as-grown CoAPO-5 crystals with increasing the cobalt molar ratio $x = [Co]/([Co]+[Al])$ in the starting gel, while the aluminium content in the crystals is correspondingly decreased. In contrast, the phosphorus content in the crystals kept a constant. These results strongly suggested that Al^{3+} was substituted by Co^{2+}, which lead to the generation of negatively charged framework and Brønsted acid sites.

The vibration spectra provide useful information about the organic template molecules inside the channels of the as-grown CoAPO-5 crystals. Micro-Raman spectra of TPA molecules inside the channels of the CoAPO-5 crystal in the C−H bond-stretching region $(2700 \sim 3150\,cm^{-1})$ measured at ambient condition are shown in Fig. 6.2. The spectra of liquid TPA and the aqueous solution of tripropylammonium fluoride are also shown in the figure as a reference. From this figure, we can observe that the C−H bond-stretching modes of the TPA inside the channels of the CoAPO-5 single crystal at 2890, 2948, and $2975\,cm^{-1}$ obviously shifted

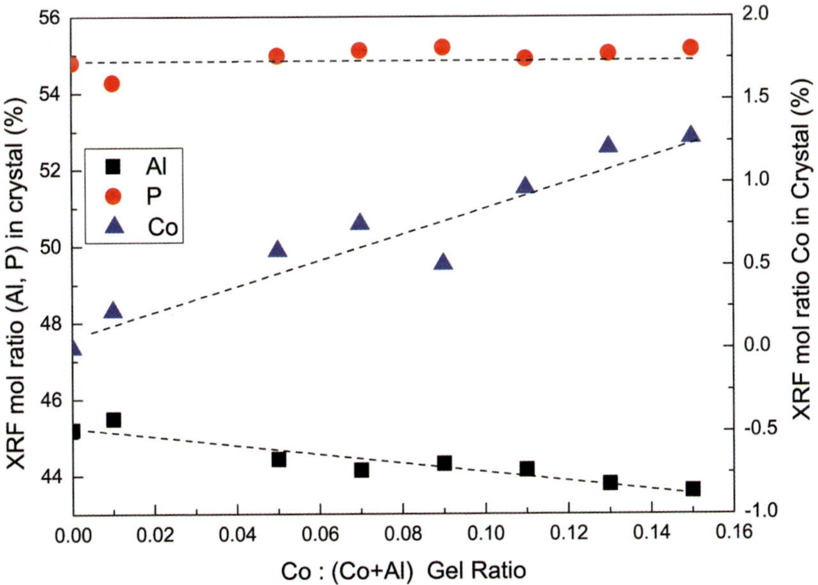

Fig. 6.1 The molar ratios of cobalt, aluminium, and phosphorus measured by the X-ray fluorescence element analyzer

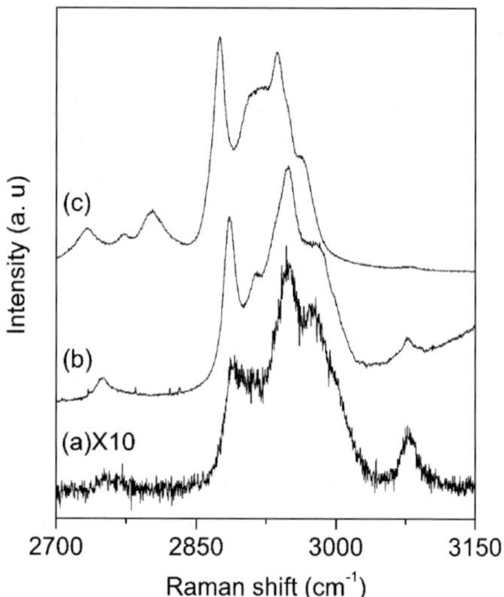

Fig. 6.2 Raman spectra of: (a) CoAPO-5 single crystal with TPA inside the channels, (b) tripropylammonium fluoride, aqueous solution, and (c) liquid TPA

to a higher frequency in comparison with that of liquid-phase TPA. The corresponding shift in the CH_3 anti-symmetrical deformation mode at $1455\,cm^{-1}$ (not shown) has also been observed. The Raman-active vibration frequencies of the TPA molecules in the CoAPO-5 crystals channels are close to those of the tripropylammonium fluoride. Thus, it can be expected that the TPA molecules are protonated inside the channels of CoAPO-5 single crystal. The TPA molecules in the as-grown CoAPO-5 crystal channels may have three predominating forms: (1) tripropylammonium hydroxide ($(CH_3CH_2CH_2)_3NH^+OH^-$), (2) tripropylammonium fluoride ($(CH_3CH_2CH_2)_3NH^+F^-$) when HF is used to restrain the nucleation and slow down the crystallization during the crystal growth, and (3) tripropylammonium cation compensating of the negatively charged framework.

A mass spectrometer was used to in-situ monitor the pyrolysis process of the TPA molecules accommodated in the CoAPO-5 crystals. The mass spectra recorded at temperatures ranging from 373 to 623 K are shown in Fig. 6.3. By elevating temperature from 300 K up to 473 K, the signal intensity of water molecules ($m/z = 18$ and 17) converted from the tripropylammonium hydroxide was increased gradually. Then the signals at $m/z=114$, 86, 72, 41 and 30 resulted from the neutral $(CH_3CH_2CH_2)_3N$ molecules were observed. The intensity of these signals increased when the sample was heated to 623 K, implying the conversion of $(CH_3CH_2CH_2)_3NH^+OH^-$ to $(CH_3CH_2CH_2)_3N$ occurred in the temperature region of $473\sim648$ K:

$$(CH_3CH_2CH_2)_3NH^+OH^- \rightarrow (CH_3CH_2CH_2)_3N + H_2O, \ (373-473K)$$

It is worth pointing out that the water signals decreased in the temperature range $473\sim598$K, meanwhile the signal intensity of TPA increased. This result can be attributed to the fact that the occluded tripropylammonium fluoride is more stable than the hydroxide, which is comparable to the results of Soulard et al. (1987) who

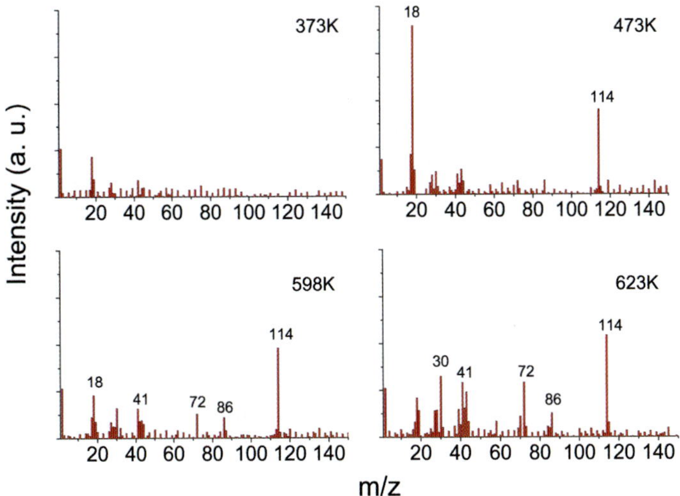

Fig. 6.3 Mass spectra obtained at various points during the thermal pyrolysis of template occluded within CoAPO-5 crystals in the temperature range 373 ∼ 623 K

investigated decomposition of tetrapropylammonium ions in MFI-type zeolites by means of thermoanalytical techniques.

$$(CH_3CH_2CH_2)_3NH^+F^- \rightarrow (CH_3CH_2CH_2)_2N + HF, \ (598{-}648\ K).$$

Figure 6.4 shows the mass spectra of the carbon precursors measured in a higher temperature region (648–823K). The signal intensity of the TPA molecules started

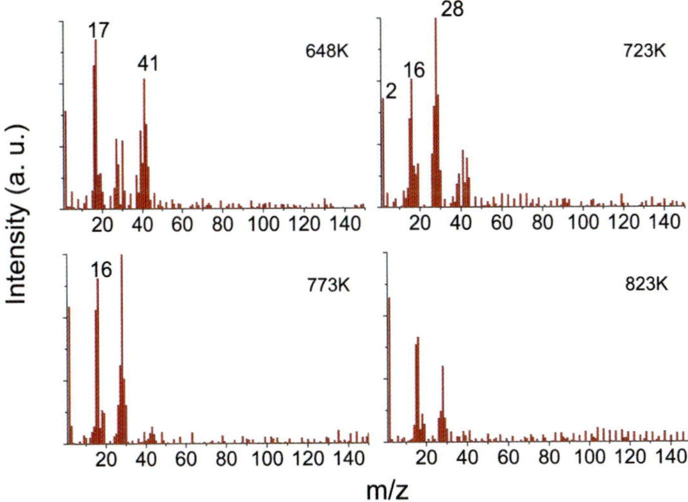

Fig. 6.4 Mass spectra obtained at various points during the thermal pyrolysis of template occluded within CoAPO-5 crystals in the temperature range 648 ∼ 823 K

to decrease when the temperature is increased to 648 K, meanwhile, the signals of propylene at $m/z = 42$, 41, 39 and 27 and ammonia (NH_3) at $m/z = 17$ and 16 were observed. The carbon precursor molecules were decomposed stepwisely into lighter molecules through sequential abstraction of propylene. It is interesting to note that, no signals of $(CH_3CH_2CH_2)_2NH$ and $(CH_3CH_2CH_2)NH_2$ were observed in these mass spectra, although the signals of propylene and ammonia molecules were clearly seen. This is due to the fact that the $(CH_3CH_2CH_2)_3NH^+$ cations are strongly attracted by the negative charged lattice (Z^-). The subsequent pyrolysis reaction of the $(CH_3CH_2CH_2)_3NH^+Z^-$ can be described by the following reaction equations:

$$(CH_3CH_2CH_2)_3NH^+Z^- \rightarrow (CH_3CH_2CH_2)_2NH^+Z^- + C_3H_6, \ (623\text{–}673 \text{ K})$$
$$(CH_3CH_2CH_2)_2NH_2^+Z^- \rightarrow (CH_3CH_2CH_2)_2NH_3^+Z^- + C_3H_6,$$
$$(CH_3CH_2CH_2)NH_3^+Z^- \rightarrow NH_3 + C_3H_6 + H^+Z^-.$$

A similar mechanism has been reported by Nowotny et al. (1991) and Park et al. (1984) who investigated the decomposition of tetrapropylammonium cations inside the channels of ZSM-5 crystals. The decomposition of the $(CH_3CH_2CH_2)_3NH^+Z^-$ composites leads to a successive release of C_3H_6 molecules. These C_3H_6 molecules are pyrolyzed into smaller molecules (such as CH_4 and C_2H_4) in the channels. This was evidenced by the signals at $m/z = 28$, 27, 26 attributable to C_2H_4; and the signals at $m/z = 16$, 15 attributable to CH_4. These small hydrocarbon molecules were subsequently carbonized. As an evidence, the intensity of H_2 at $m/z = 2$ was increased. All signals became weaker and finally undetectable when the sample was heated to 823 K, indicating a complete carbonization of hydrocarbon precursor.

The carbonization process of the TPA molecules inside the channels of CoAPO-5 crystals was further analyzed by micro-Raman spectroscopy. Raman spectra of the TPA@CoAPO-5 crystals measured at various temperatures from 373 to 773 K in a vacuum of 10^{-3} mbar are shown in Fig. 6.5. When the CoAPO-5 crystals were treated at a temperature below 573 K, the spectrum remained essentially the same as that at 373 K, except that the intensities decreased. It shows typical characteristic Raman-active modes of TPA molecules: The peaks near 2850–3050 cm^{-1} are assigned to the CH_3 and CH_2 symmetric and anti-symmetric stretching modes, the peak at about 1460 cm^{-1} is due to the CH_3 anti-symmetrical deformation mode and that near $967 \sim 1150 \ cm^{-1}$ is attributed to $C - N$ stretching mode (During, 1989). When the sample was pyrolyzed at temperature 623 K for 3 hours, all signals related to TPA molecules disappeared from the spectrum, indicating that the carbon precursor TPA molecules start to decompose at this temperature. At even higher temperature, the G-band at 1600 cm^{-1} appeared when the sample was heated at temperature above 648 K, implying that carbon precursor TPA molecules was carbonized inside the channels of CoAPO-5 single crystal. With still increasing temperature, the relative intensity of G-band gradually strengthened. The sharpness of the peaks is an indicator that carbon atoms are highly ordered in the channels. While the radial breathing mode (RBM) frequencies at 510 and 550 cm^{-1} (Li et al., 2004a) and the disorder-induced D line appear at temperature up to

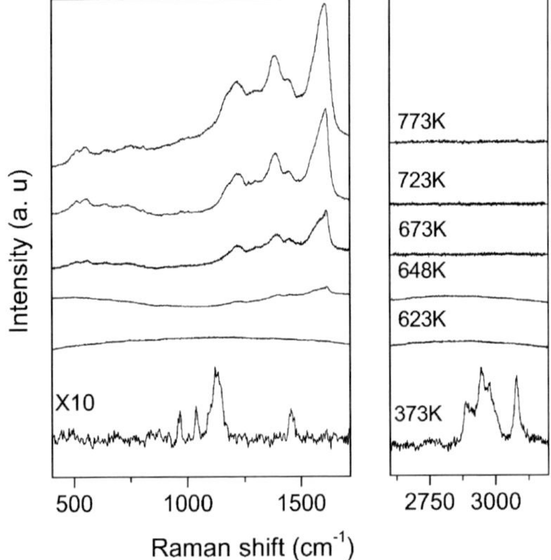

Fig. 6.5 Raman spectra of the TPA contained inside the channels of CoAPO-5 crystal were measured at various temperatures in a vacuum of 10^{-3} mbar

673 K, an indicative of forming the 0.4 nm carbon nanotubes at the temperature. As a result, the crystals turned homogeneous black with strong optical anisotropy (Li et al., 2001). The relative intensity of RBM is increased with the increasing temperature. All these results indicate that the 0.4 nm SWNTs inside the channels of CoAPO-5 crystal started to be formed from small graphite sheet. Compared with that of $AlPO_4$-5 single crystal, the temperature of TPA carbonization inside the channels of CoAPO-5 single crystal is notably decreased (Sun et al., 1999). It should be pointed out that there also exist Raman signals for the CoAPO-5 lattice at frequencies of $450 \, cm^{-1}$, due to the P-O-Al vibration (Holmes et al., 1994). However, the Raman intensity of SWNTs is at least 20 times stronger than that of CoAPO-5, owing to the resonant behavior. Thus the contribution of the vibrations from the CoAPO-5 framework to the Raman spectra is negligible (Zhai et al., 2006c).

6.4 Conclusions

In summary, the identification of the volatiles and the characterization of the intermediate materials during pyrolysis of TPA carbon precursor molecules trapped in as-grown CoAPO-5 crystals have been studied. The present data supported the hypothesis that the TPA exists as tripropylammonium ions in crystals channels and latter decomposes into dipropyl- and propylammonium cations by sequential abstraction of propylene, in the temperature range 423–648 K. The result of micro-Raman showed that the 0.4 nm nanotubes were formed at temperature about 673 K. Compared with that of $AlPO_4$-5 single crystal, the temperature of TPA carbonization inside the channels of CoAPO-5 single crystal is notably decreased because of the catalyst effect of Co-ions in the AFI framework.

This means that a small device has a maximum conductance of

$$\frac{dI}{dV} \sim \frac{e\gamma_1/2\hbar}{2\gamma_1} = \frac{e^2}{4\hbar}$$

This is not too far from the correct result

$$G_{\max} = e^2/h \approx 25.8\ K\Omega \tag{7.9}$$

which is one of the seminal results of mesoscopic physics that was not known before 1988. One could view this as a consequence of the energy broadening required by the uncertainty principle which comes out automatically in the full Landauer-NEGF approach (Fig. 7.3b), but has to be inserted by hand into the simpler version we are using where the density of states $D(E)$ is an input parameter (Fig. 7.3a).

For our examples in this paper we will use a density of states that is constant in the energy range of interest, for which the current-voltage characteristic is basically linear.

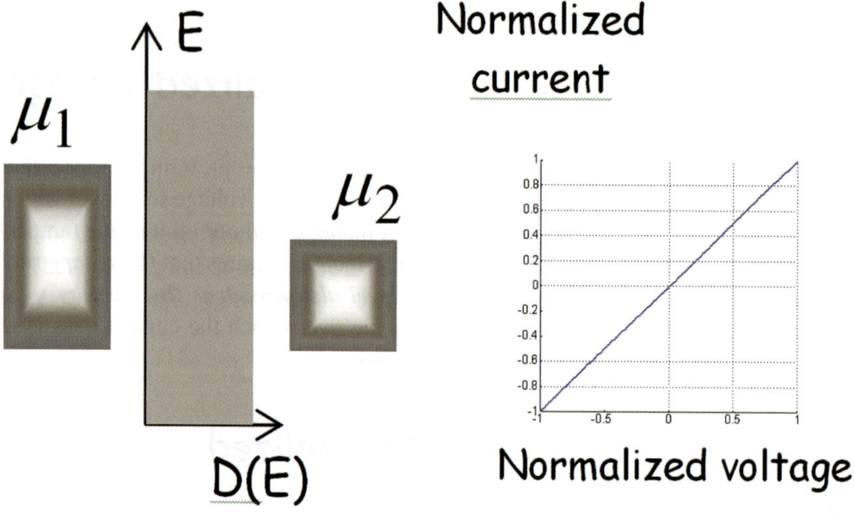

7.2.3 Current-Voltage Relation: With Demons

Defining $i_s(E)$ as the "scattering current" *induced by interaction of the electrons with the "demon"*, we can write

$$i_1(E) = i_2(E) + i_s(E) \tag{7.10}$$

This current can be modelled in general as a difference between two processes one involving a loss of energy ε from the demon and the other involving a gain of energy by the demon.

$$i_s(E) = (e/\hbar)\, \gamma_s\, D(E) \int d\varepsilon\, D(E + \varepsilon)$$
$$[F(\varepsilon)\, f(E)\, (1 - f(E + \varepsilon))$$
$$- F(-\varepsilon) f(E + \varepsilon)\, (1 - f(E))] \tag{7.11}$$

The basic principle of equilibrium statistical mechanics requires that if the demon is in equilibrium at some temperature T_D then the strength $F(\varepsilon)$ of the energy loss processes is related to the strength $F(-\varepsilon)$ of energy gain processes by the ratio:

$$F(\varepsilon) = F(-\varepsilon)\, \exp\left(-\varepsilon/k_B T_D\right) \tag{7.12}$$

With a little algebra one can show that this relation ensures that if the electron distribution $f(E)$ is given by an equilibrium Fermi function with the same temperature T the two terms in Eq. (7.11) will cancel out. This result is independent of the detailed shape of the function $F(\varepsilon)$ describing the spectrum of the demon, as long as Eq. (7.9) is true. This means that *if the demon is in equilibrium with the electrons with the same temperature, there can be no net flow of energy either to or from the demon*. Indeed one could view this as the basic principle of equilibrium statistical mechanics and work backwards to obtain Eq. (7.12) as the condition needed to ensure compliance with this principle.

To summarize, if the electrons in the channel do not interact with any "demons", the current voltage characteristics are obtained from Eq. (7.7) using the Fermi functions from Eq. (7.5). For the more interesting case with interactions, we solve for the distribution $f(E)$ inside the channel from Eqs. (7.4), (7.10) and (7.11) and then calculate the currents. Usually the current flow is driven by an external voltage that separates the electrochemical potentials μ_1 and μ_2 in Eq. (7.7). But thermoelectric currents driven by a difference in temperatures T_1 and T_2 can also be calculated from this model (Paulsson and Datta, 2003) as mentioned in the introduction. Our focus here is on a different possibility for energy conversion, namely through out-of-equilibrium demons.

7.2.4 *Where is the Heat?*

Before we move on, let me say a few words about an important conceptual issue that caused much argument in the early days: *Where is the heat dissipated in a ballistic conductor?* After all, if there is no demon to take up the energy, there cannot be any dissipation inside the channel. The answer is that the transiting electron appears as a hot electron in the drain (right) contact and leaves behind a hot hole in the source (left) contact (see Fig. 7.4). The contacts are immediately restored to their equilibrium states by unspecified dissipative processes operative within each contact. These processes can be quite complicated but are usually incorporated surreptitiously into the model through what appears to be an innocent boundary condition, namely that the electrons in the contacts are always maintained in thermal equilibrium described by Fermi distributions (Eq. (7.5)) with electrochemical potentials μ_1 and μ_2 and temperatures T_1 and T_2.

To understand the spatial distribution of the dissipated energy it is useful to look at the energy current which is obtained by replacing the charge 'e' with the energy E of the electron:

$$I_{E_1} = (1/e) \int dE \; E \; i_1(E), \; I_{E_2} = (1/e) \int dE \; E \; i_2(E) \qquad (7.13)$$

The energy currents at the source and drain contacts are written simply as

$$I_{E_s}(E) \; = \; (\mu_1/e) \; I_1, \; I_{E_d}(E) \; = \; (\mu_2/e) \; I_2 \qquad (7.14)$$

assuming that the entire current flows around the corresponding electrochemical potentials.

Figure 7.4 shows a spatial plot of the energy current from the source end to the drain end for a uniform 1-D ballistic conductor with a voltage of 50 mV

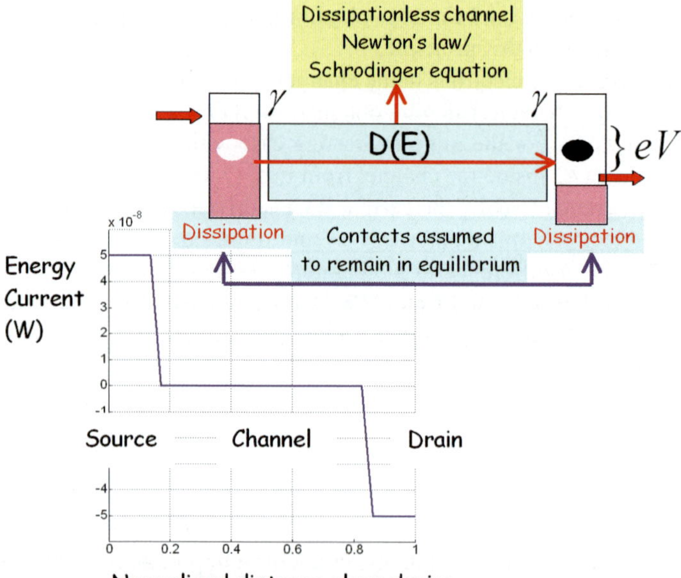

Fig. 7.4 Plot of energy current in a 1-D ballistic conductor with $G = e^2/h \approx 40 \, \mu S$ and an applied voltage of $V = 0.05$ volts. Energy dissipated is given by the drop in the energy current, showing that the Joule heating $V^2 G = 0.1 \, \mu W$ is divided equally between the two contact-channel interfaces

applied across it. For a conductor with no demon for electrons to exchange energy with, $i_1(E) = i_2(E)$ making the energy current uniform across the entire channel implying that no energy is dissipated inside the channel. But the energy current entering the source contact is larger than this value while that leaving the drain contact is lower. Wherever the energy current drops, it means that the rate at which energy flows in is greater than the rate at which it flows out, indicating a net energy dissipation. Clearly in this example, $0.05\,\mu W$ is dissipated in each of the two contacts thus accounting for the expected Joule heating given by $V^2 G = 0.1\,\mu W$.

Real conductors have distributed demons throughout the channel so that dissipation occurs not just in the contacts but in the channel as well. Indeed we commonly assume the Joule heating to occur uniformly across a conductor. But there are now experimental examples of nanoscale conductors that would have been destroyed if all the heat were dissipated internally and it is believed that the conductors survive only because most of the heat is dissipated in the contacts which are large enough to get rid of it efficiently. The idealized model depicted in Fig. 7.4 thus may not be too far from real nanodevices of today.

As I mentioned in the introduction, what distinguishes the Landauer model (Fig. 7.4) is the physical separation of dynamics and dissipation clearly showing that what makes transport an irreversible process is the continual restoration of the contacts back to equilibrium. Without this ***repeated restoration***, all flow would cease once a sufficient number of electrons have transferred from the source to the drain. Over a 100 years ago Boltzmann showed how pure Newtonian dynamics could be supplemented to describe transport processes, and his celebrated equation stands today as the centerpiece for the flow of all kinds of particles. Boltzmann's approach too involved "repeated restoration" through an assumption referred to as the "Stohsslansatz" (see for example, McQuarrie, 1976) Today's devices often involve Schrodinger dynamics in place of Newtonian dynamics and the non-equilibrium Green function (NEGF) method that we use (Fig. 7.3b) supplements the Schrodinger equation with similar assumptions about the repeated restoration of the surroundings that enter the evaluation of the various self-energy functions Σ or the corresponding quantities γ in the simpler model that we are using.

"Contacts" and "demons" are an integral part of all devices, the most common demon being the phonon bath for which the relation in Eq. (7.12) is satisfied by requiring that energy loss $\sim N$ and energy gain $\sim (N + 1)$, N being the number of phonons given by the Bose Einstein factor if the bath is maintained in equilibrium. Typically such demons add channels for dissipation, but our purpose here is to show how suitably engineered out-of-equilibrium demons can act as sources of energy.

For this purpose, it is convenient to study a device specially designed to accentuate the impact of the demon. Usually the interactions with the demon do not have any clear distinctive effect on the terminal current that can be easily detected. But in this special device, ideally no current flows ***unless the channel electrons interact with the demon***. Let me now describe how such a device can be engineered.

7.3 Elastic Demon

Let us start with a simple 1-D ballistic device but having two rather special kinds of contacts. The source contact allows one type of spin, say "black" (drawn pointing to the right), to go in and out of the channel much more easily than the other type, say "white" (drawn pointing to the left, Fig. 7.5a). Devices like this are called spin-valves and are widely used to detect magnetic fields in magnetic memory devices (see for example the articles in Heinreich and Bland, 2004).

Although today's spin-valves operate with contacts that are far from perfect, since we are only trying to make a conceptual point, let us simplify things by assuming contacts that are perfect in their discrimination between the two spins. One only lets black spins while the other only lets white spins to go in and out of the channel. We then have the situation shown in Fig. 7.5b and no current can flow since neither black nor white spins communicate with both contacts. But if we introduce impurities into the channel (the demon) with which electrons can interact and flip their spin, then current flow should be possible as indicated: black spins come in from the left, interact with impurities to flip into white spins and go out the right contact (Fig. 7.5c).

Consider now what happens if the impurities are say all white (Fig. 7.6a). Electrons can now flow only when the bias is such that the source injects and drain collects (positive drain voltage), but not if the drain injects and the source collects (negative drain voltage). This is because the source injects black spins which interact with the white impurities, flip into whites and exit through the drain. But the drain only injects white spins which cannot interact with the white impurities and cannot cross over into the source. Similarly if the impurities are all black, current flows

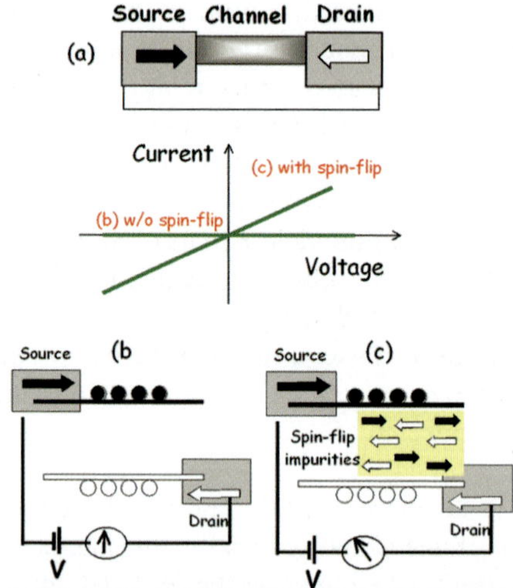

Fig. 7.5 Anti-parallel (AP) Spin-valve: (a) Physical structure, (b) no current flows if the contacts can discriminate between the two spins perfectly, (c) current flow is possible if impurities are present to induce spin-flip processes

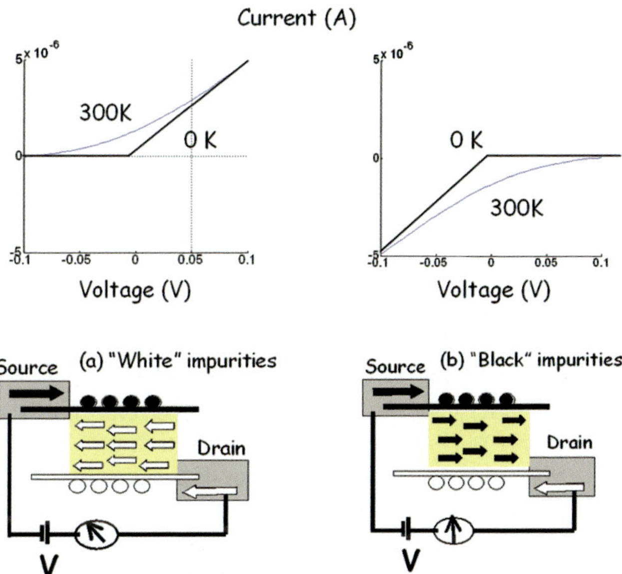

Fig. 7.6 Perfect AP Spin-valve with (a) white (down spin, drawn as pointing to the left) impurities and (b) black (up spin, drawn as pointing to the right) impurities

only for negative drain voltage (Fig. 7.6b). At non-zero temperatures the cusps in the current-voltage characteristics get smoothed out and we get the smoother curves shown in Fig. 7.6 a, b.

Note the surprising feature of the plots at $T = 300$ K: *there is a non-zero current even at zero voltage*! This I believe is correct. Devices like those shown in Fig. 7.6 a, b with polarized impurities could indeed be used to generate power and one could view the system of impurities as a Maxwell's demon that lets electrons go preferentially from source to drain or from drain to source. The second law is in no danger, since the energy is extracted at the expense of the entropy of the system of impurities that collectively constitute the demon. Assuming the spins are all non-interacting and it takes no energy to flip one, the polarized state of the demon represents an unnatural low entropy state. Every time an electron goes through, a spin gets flipped raising the entropy and the flow will eventually stop when demon has been restored to its natural unpolarized state with the highest entropy of $N_I k_B \ln 2$, where N_I is the number of impurities. To perpetuate the flow an external source will have to spend the energy needed to maintain the demon in its low entropy state.

7.3.1 Current Versus Voltage: Summary of Equations

Let me now summarize the equations that I am using to analyze structures like the one in Fig. 7.6 quantitatively. Basically it is an extension of the equations described earlier (see Eq. (7.6), Section 7.2) to include the two spin channels denoted by the subscripts "u" (up or black) and "d" (down or white):

$$i_{1,u}(E) = (e/\hbar)\gamma_{1,u}\, D_u(E)\, [f_1(E) - f_u(E)]$$
$$i_{1,d}(E) = (e/\hbar)\gamma_{1,d}\, D_d(E)\, [f_1(E) - f_d(E)] \tag{7.15}$$
$$i_{2,u}(E) = (e/\hbar)\gamma_{2,u}\, D_u(E)\, [f_u(E) - f_2(E)]$$
$$i_{2,d}(E) = (e/\hbar)\, \gamma_{2,d}\, D_d(E)\, [f_d(E) - f_2(E)] \tag{7.16}$$

For the scattering current caused by the demon we write (see Eq. (7.11))

$$i_{s,u}(E) = -\, i_{s,d}(E) = (e/\hbar)\, (2\pi\, J^2 N_I)\, D_u(E)\, D_d(E)$$
$$[F_d\, f_u(E)\, (1 - f_d(E)) - F_u\, f_d(E)\, (1 - f_u(E))] \tag{7.17}$$

noting that we are considering an elastic demon that can neither absorb nor give up energy ($\varepsilon = 0$). The first term within parenthesis in Eq. (7.17) represents an up electron flipping down by interacting with a down-impurity while the second term represents a down electron flipping up by interacting with an up-impurity. The strengths of the two processes are proportional to the fractions F_d and F_u of down and up-spin impurities. The overall strength of the interaction is governed by the number of impurities N_I and the square of the matrix element J governing the electron-impurity interaction.

Equation (7.16) can be visualized in terms of an equivalent circuit (Fig. 7.7) if we think of the various f's as "voltages" since the currents are proportional to differences in 'f' just as we expect for conductances defined as

$$g_{1,u} = (e^2/\hbar)\, \gamma_{1,u}\, D_u\ etc.$$

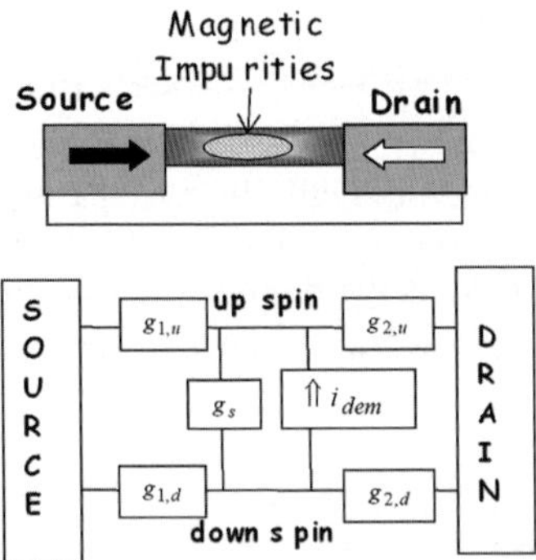

Fig. 7.7 An anti-parallel (AP) spin-valve with spin-flip impurities and a simple equivalent circuit to help visualize the equations we use to describe it, namely Eqs. (7.15) through (7.17) (adapted from Datta, 2005 b,c)

The scattering current (Eq. (7.17)) too could be represented with a conductance g_s if we set $F_d = F_u = 0.5$

$$i_s = (e/\hbar)(\pi J^2 N_I) D_u(E) D_d(E)(f_d - f_u) \qquad (7.18a)$$

However, this is true only if the impurities are in equilibrium, while the interesting current-voltage characteristics shown in Fig. 7.6 require an out-of-equilibrium demon with $F_d \neq F_u$. So we write the total scattering current from Eq. (7.17) as a sum of two components, one given by Eq. (7.18a) and another proportional to $(F_d - F_u)$ which we denote with a subscript 'dem':

$$i_{dem} = (F_d - F_u) N_I / \tau_0$$

$$\frac{1}{\tau_0} = (e/\hbar)\,(\pi J^2)\, D_u(E) D_d(E)\, [f_d(1 - f_u) + f_u(1 - f_d)] \qquad (7.18b)$$

Equation (7.16) and (7.17) can be solved to obtain the distribution functions $f_u(E)$ and $f_d(E)$ by imposing the requirement of overall current conservation (cf. Eq. (7.11)):

$$i_{1,u}(E) = i_{2,u}(E) + i_{s,u}(E)$$
$$i_{1,d}(E) = i_{2,d}(E) + i_{s,d}(E) \qquad (7.19)$$

The currents are then calculated and integrated over energy to obtain the terminal currents shown in Fig. 7.6. We can also find the energy currents using equations like Eqs. (7.14) and (7.15) and the results are shown in Fig. 7.8 for $F_d - F_u = -1$ and for $F_d - F_u = 0$ each with a voltage difference of $2k_B T = 50\,\mathrm{mV}$ between the two terminals. With $F_d - F_u = 0$ the direction of current is in keeping with an external battery driving the device. But with $F_d - F_u = -1$, the external current flows against the terminal voltage indicating that the device is acting as a source of energy driving a load as shown in the inset. This is also borne out by the energy current flow which shows a step up at each interface indicating that *energy is being absorbed from the contacts* ($\sim 10\,\mathrm{nW}$ from each) and delivered to the external load.

But isn't this exactly what the second law forbids? After all if we could just absorb energy from our surroundings (the contacts) and do useful things, there would be no energy problem. The reason this device is able to perform this impossible feat is that the impurities are assumed to be held in a non-equilibrium state with very low entropy. A collection of N_I impurities can be unpolarized in 2^{N_I} different ways having an entropy of $S = N_I k_B \ln 2$. But it can be completely polarized ($F_d - F_u = \pm 1$) in only one way with an entropy of $S = 0$. What this device does is to exchange entropy for energy. Many believe that the universe too is evolving the same way, with constant total energy, from a particularly low entropy state continually towards a higher entropy one. But that is a different matter.

To have our device continue delivering energy indefinitely we will need an external source to maintain the impurities in their low entropy state which will cost

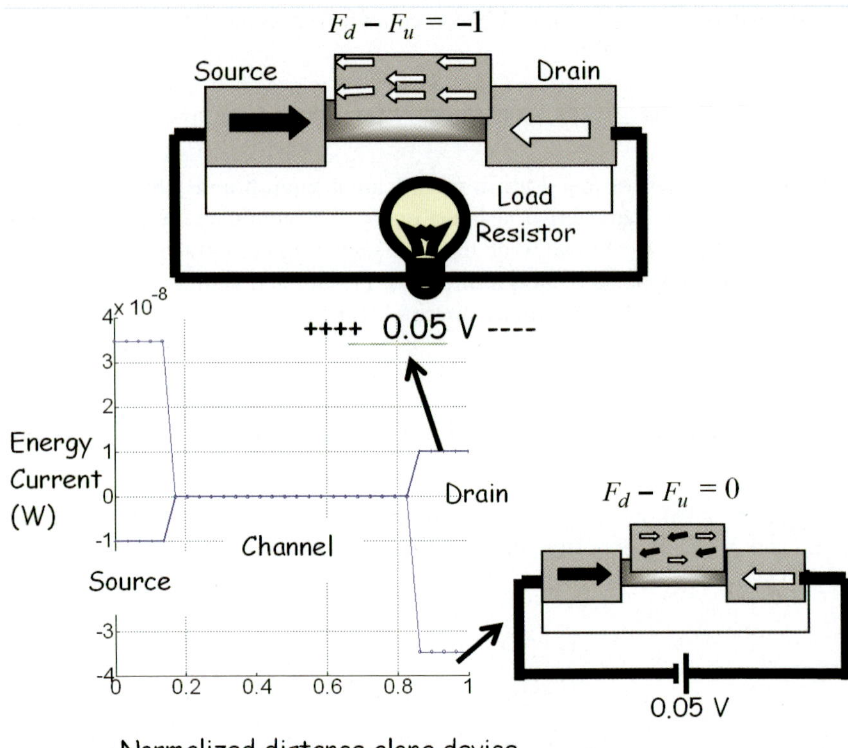

Fig. 7.8 Energy currents with $F_d - F_u = -1$ and with $F_d - F_u = 0$ each with a voltage difference of $2k_BT = 50\,\text{mV}$ between the two terminals. With $F_d - F_u = 0$ the direction of current is in keeping with an external battery driving the device. But with $F_d - F_u = -1$, the external current flows against the terminal voltage indicating that the device is acting as a source of energy driving a load as shown in the inset. This is also borne out by the energy current flow which shows a step up at each interface indicating that energy is being absorbed from the contacts ($\sim 10\,\text{nW}$ from each) and delivered to the external load

energy. The details will depend on the actual mechanism used for the purpose but we will not go into this. Note that if we do not have such a mechanism, the current will die out as the spins get unpolarized. This depolarization process can be described by an equation of the form:

$$dP_I/dt + \left(\frac{1}{\tau_I} + \frac{1}{\tau_0}\right) P_I = i_s \tag{7.20}$$

where τ_0 and i_s are related to the scattering current as defined in Eqs. (7.18), while the additional time constant τ_I represents processes unrelated to the channel electrons by which impurities can relax their spins.

7.4 Inelastic Demon

We have argued above that although one can extract energy from polarized impurities, energy is needed to keep them in that state since their natural high entropy state is the unpolarized one. It would seem that one way to keep the spins naturally in a polarized state is to use a nanomagnet, a collection of spins driven by a ferromagnetic interaction that keeps them all pointed in the same direction. Could such a magnet remain polarized naturally and enable us to extract energy from the contacts forever? The answer can be "yes" if the magnet is at a different temperature from the electrons. What we then have is a heat engine working between two reservoirs (the electrons and the magnet) at different temperatures and we will show that its operation is in keeping with Carnot's principle as required by the second Law.

To model the interaction of the electrons with the nanomagnet we need to modify the expression for the scattering current (Eq. (7.17)) for it now takes energy to flip a spin. We can write

$$i_{s,u}(E) = -\, i_{s,d}(E) \;=\; (e/\hbar)\,(2\pi\, J^2 N_I) \int d\varepsilon\; D_u(E)\; D_d(E+\varepsilon)$$

$$[F(\varepsilon)\, f_u(E)\,(1-f_d(E+\varepsilon)) \;-\; F(-\varepsilon)\, f_d(E+\varepsilon)\,(1-f_u(E))] \tag{7.21}$$

where $F(\varepsilon)$ denotes the magnon spectrum obeying the general law stated earlier (see Eq. (7.13)) if the magnet has a temperature T_D. Equation (7.21) can be solved along with Eqs. (7.15), (7.16) and (7.19) as before to obtain currents, energy currents etc. But let us first try to get some insight using simple approximations.

If we assume that the electron distribution functions $f_u(E)$ and $f_d(E)$ are described by Fermi functions with electrochemical potentials μ_u and μ_d respectively and temperature T, and make use of Eq. (7.13) we can rewrite the scattering current from Eq. (7.21) in the form

$$i_{s,u}(E) = -i_{s,d}(E) = (e/\hbar)(2\pi J^2 N_1) \int d\varepsilon\, D_u(E) D_d(E+\varepsilon)$$

$$F(\varepsilon) f_u(E)(1-f_d(E+\varepsilon)) \left(1 - \frac{F(-\varepsilon)}{F(+\varepsilon)}\,\frac{f_d(E+\varepsilon)}{1-f_d(E+\varepsilon)}\,\frac{1-f_u(E)}{f_u(E)}\right)$$

$$= (e/\hbar)(2\pi J^2 N_1) \int d\varepsilon\, D_u(E) D_d(E+\varepsilon) F(\varepsilon) f_u(E)(1-f_d(E+\varepsilon))$$

$$\left[1 - \exp\left(\frac{\varepsilon}{k_B T_D} + \frac{\mu_d - \varepsilon - \mu_u}{k_B T}\right)\right] \tag{7.22}$$

If we further assume the exponent to be small, so that $1 - \exp(-x) \approx x$, we can write

$$i_{s,u} = -i_{s,d} \sim \left[\frac{\mu_u - \mu_d}{k_B T}\right] + \left[\frac{\varepsilon}{k_B T} - \frac{\varepsilon}{k_B T_D}\right] \tag{7.23}$$

The first term represents a dissipative current proportional to the potential difference and can be represented by a conductance like the g_s in Fig. 7.7, while the second term isg the demon induced source term that can be harnessed to do external work. It vanishes when the demon temperature T_D equals the electron temperature T.

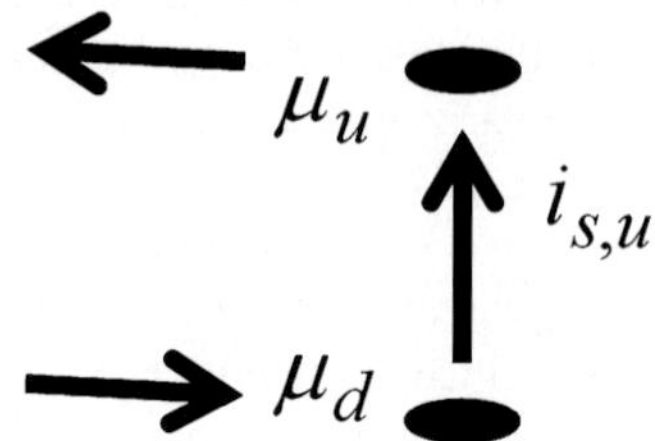

To be specific, let us assume that the demon is cooler than the rest of the device $(T_D < T)$ so that

$$\frac{\varepsilon}{k_B T} - \frac{\varepsilon}{k_B T_d} < 0$$

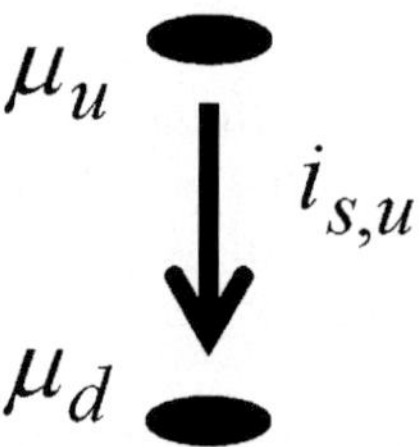

giving rise to a flow of electrons out of the up spin node and back into the downspin node. If we use this to drive an external load then $\mu_u - \mu_d > 0$, which will tend to reduce the net current given by Eq. (7.23) and the maximum output voltage one can get corresponds to "open circuit" conditions with zero current:

$$\frac{\mu_u - \mu_d}{k_B T} \leq \frac{\varepsilon}{k_B T_D} - \frac{\varepsilon}{k_B T} \tag{7.24}$$

This expression has a simple interpretation in terms of the Carnot principle. Every time an electron flows around the circuit giving up energy ε to the demon, it delivers energy $\mu_u - \mu_d$ to the load. All this energy $\mu_u - \mu_d + \varepsilon$ is absorbed from the contacts and the Carnot principle requires that

$$\text{Energy from contacts} / T \leq \text{Energy to demon}/T_D$$

that is,

$$\frac{\mu_u - \mu_d + \varepsilon}{T} \leq \frac{\varepsilon}{T_D}$$

which is just a restatement of Eq. (7.24). Note that usual derivations of the Carnot principle do not put the system simultaneously in contact with two reservoirs at

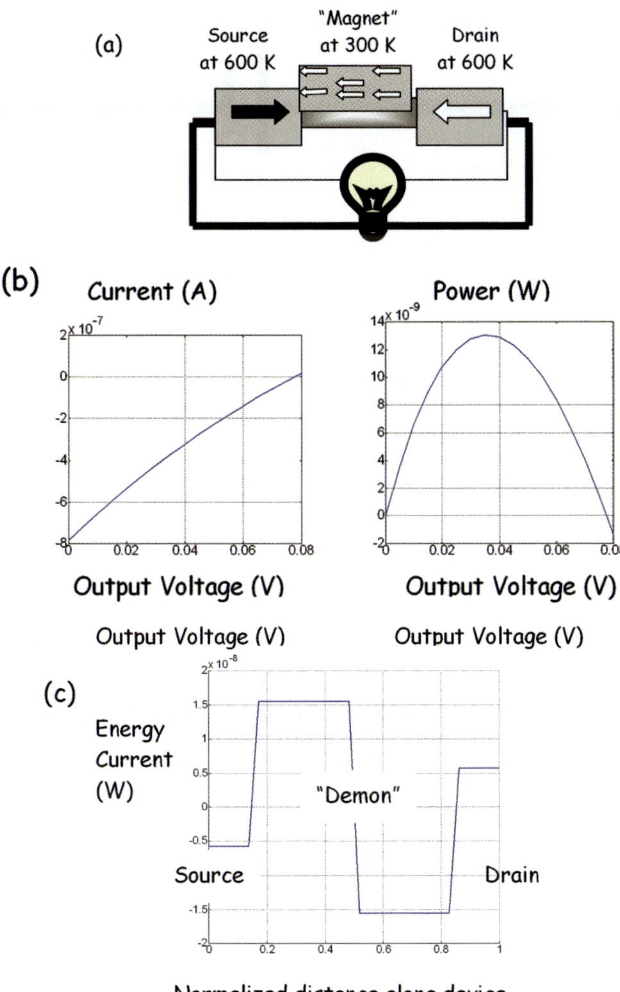

Fig. 7.9 Heat engine: (a) AP Spin-valve with a cooled magnet as Maxwell's demon controlling the flow of electrons. (b) Output current and output power versus output voltage as the load is varied from short circuit ($V = 0$) to open circuit conditions ($I = 0$). The spectrum of the magnet is assumed to consist of a single energy $\varepsilon = 2k_B T = 50\,\text{meV}$. (c) Energy current profile assuming a load such that the output voltage is 50 mV

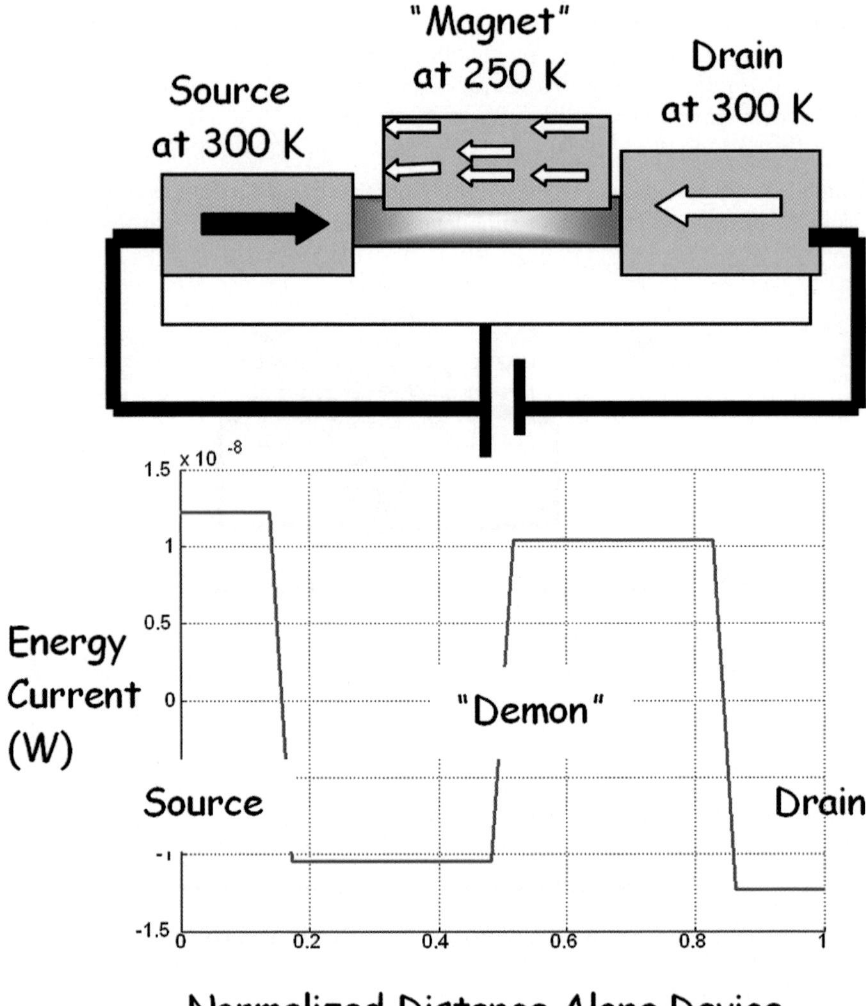

Fig. 7.10 (a) Heat engine from Fig. 7.9 operated as a refrigerator by applying an external battery to inject downspin (white) electrons from the drain that flip down the thermally created up spins in the magnet, thus cooling it. (b) Energy current profile showing that energy is absorbed from the external battery and the demon and dissipated in the source and drain contacts

different temperatures as we have done. Our treatment is closer in spirit to the classic discussion of the ratchet and pawl in the Feynman Lectures (Feynman et al., 1963).

Figure 7.9 shows numerical results obtained by solving Eqs. (7.21), (7.19), (7.15) and (7.16). The extracted power is a maximum (Fig. 7.9d) when the output voltage is somewhere halfway between zero and the maximum output voltage of 80 mV (a few $k_B T$). Figure 7.9b shows the energy current profile at an output voltage of 50 mV: energy is absorbed from the source and drain contacts and given up partly to the demon and the rest is delivered to the external load. The efficiency (energy given to load/energy absorbed from contact) is a maximum close to open circuit conditions, but the energy delivered is very small at that point as evident from Fig. 7.9d.

As one might expect, one can also operate the same device as a refrigerator by using an external source that seeks to inject downspins (white) from the drain contact that flip back thermally created up spins in the magnet thus cooling it. It is evident from the energy current profile shown in Fig. 7.10 that in this case, energy is absorbed from the demon and from the battery and given up to the source and drain contacts.

7.5 Entangled Demon

In this talk I have tried to introduce a simple transparent model showing how out-of-equilibrium demons suitably incorporated into nanodevices can achieve energy conversion. At the same time this model illustrates the fundamental role played by "contacts" and "demons" in these processes. I would like to end by pointing out another aspect of contacts that I believe is important in taking us to our next level of understanding. The basic point can be appreciated by considering a simple version of the spin capacitor we started with (see Fig. 7.5) but having just one impurity (Fig. 7.11). No current can flow in this structure without spin-flip processes since the source injects black (up) electrons while the drain only collects white (down) electrons. But if the black electron interacts with the white impurity (A) to produce a black impurity then the white electron can be collected resulting in a flow of current.

The process of conversion

From   A: Black electron $\otimes$ White impurity

To   B: White electron $\otimes$ Black impurity

is incorporated into our model through the scattering current (see Eq. (7.17)). A more complete quantum transport model involving matrices (Fig. 7.2a) rather

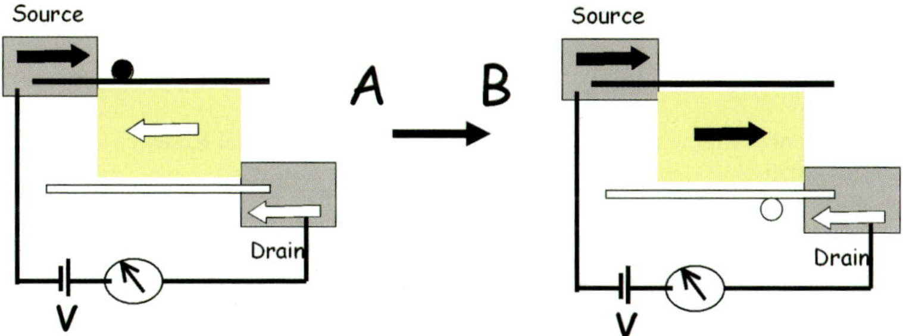

Fig. 7.11 We have described spin-flip processes in terms of a direct conversion from state A to state B. Quantum mechanics, however, requires an intermediate state consisting of a superposition of A and B before wavefunction collapse reduces it to a B. This state may have a significant role in devices with weak contacts and strong interactions in the channel, requiring a model that can deal with entangled states

than numbers (Fig. 7.2b) could be used to describe this effect, but the essential underlying assumption in either case is that the state of the electron-impurity system changes from A to B upon interaction. Quantum mechanics, however, paints a different picture of the process involving an intermediate **entangled state**. It says that the system goes from A into a state consisting of a superposition of A and B and it is only when the electron is collected by the drain that the wavefunction collapses to a B. If the collection rates $\gamma_{1,2}/\hbar$ are much larger than the interaction rate per impurity $\gamma_s/\hbar N_I$, we expect the entangled state to play a minor role. But this may not be true of devices with weak contacts and strong interactions in the channel, requiring a model that can deal with entangled states.

Entangled states are difficult to describe within the conceptual framework we have been using where both the electrons and the impurity are assumed to exist in independent states. It is hard to describe a "conditional state" where the electron is black if the impurity is white or vice versa, let alone a superposition of the two. To account for this entangled state we need to treat the electron and impurity as one big system and write rate equations for it, in the spirit of the many-electron rate equations widely used to treat Coulomb blockade but the standard approach (Beenakker, 1991; Likharev, 1999) needs to be extended to include coherences and broadening. This is an area of active research (see for example Braun et al., 2004; Braig and Brouwer, 2005) where an adequate general approach does not yet exist.

Actually correlated states (classical version of entanglement) were an issue even before the advent of quantum mechanics. Boltzmann ignored them through his assumption of "molecular chaos" or "Stohsslansatz", and it is believed that it is precisely this assumption that leads to irreversibility (see for example, McQuarrie, 1976). An intervening entangled or correlated state is characteristic of all "channel"-"contact" interactions, classical or quantum, and the increase in entropy characteristic of irreversible processes can be associated with the destruction or neglect of the correlation and/or entanglement generated by the interaction (see for example, Zhang, 2007, Datta, 2007). This aspect is largely ignored in today's transport theory, just as even the presence of a contact was barely acknowledged before the advent of mesoscopic physics. But new experiments showing the effect of entanglements on current flow are on the horizon and will hopefully lead us to the next level of understanding.

Acknowledgments This work was supported by the Office of Naval Research under grant no. N00014-06-1-0025.

References

Beenakker, C.W.J., 1991, Theory of Coulomb blockade oscillations in the conductance of a quantum dot. In Physical Review, **B44**, 1646.

Braig, S. and Brouwer, P.W., 2005, Rate equations for Coulomb blockade with ferromagnetic leads. In Physical Review, **B71**, 195324.

Braun, M., Koenig, J. and Martinek, J., 2004, Theory of transport through quantum-dot spin valves in the weak coupling regime. In Physical Review, **B70**, 195345.

Datta, S., 1989, Steady-state quantum kinetic equation. In Physical Review, **B40**, 5830.

Datta, S., 1990, A simple kinetic equation for steady-state quantum transport. In Journal of Physics: Condensed Matter, **2**, 8023.

Datta, S., 2005a, Quantum transport: Atom to transistor (Cambridge University Press).

Datta, S., 2005b, Proposal for a spin capacitor. In Applied Physics Letters, **87**, 013115.

Datta, S., 2005c, Spin dephasing and hot spins. In Proceedings of the International School of Physics "Enrico Fermi" Course CLX edited by A. D'Amico, G. Balestrino and A. Paoletti (IOS Press, Amsterdam and SIF Bologna).

Datta, S., 2006, Concepts of quantum transport, a series of video lectures, http://www.nanohub.org/courses/cqt

Datta, S., 2007, Quantum Transport as a Process of Repeated Disentanglement, Preprint.

Feynman, R.P., Leighton, R.B. and Sands, M., 1963, Lectures on Physics, vol.1, Chapter 46 (Addison-Wesley, Boston).

Fuchs, G.D., Krivotorov, I.N., Braganca, P.M., Emley, N.C., Garcia, A.G.F., Ralph, D.C., Buhrman, R.A., 2005, Adjustable spin torque in magnetic tunnel junctions with two fixed layers, In Applied Physics Letters, **86**, 152509.

Huai, Y., Pakala, M., Diao, Z., Ding, Y., 2005, Spin transfer switching current reduction in magnetic tunnel junction based dual spin filter structures. In Applied Physics Letters, **87**, 222510.

Heinreich, B. and Bland, J.A.C., (eds) 2004, Introduction to micromagnetics in ultrathin magnetic structures, vol. IV (Springer-Verlag, Berlin).

Leff, H. S. and Rex, A.F., (eds) 1990, Maxwell's demon, entropy, information and computing, Princeton Series in Physics.

Leff, H. S. and Rex, A. F., (eds) 2003, Maxwell's Demon 2: entropy, classical and quantum information, computing (IOP Publishing, Bristol).

Likharev, K., 1999, Single electron devices and their applications. In Proceedings IEEE, **87**, 606.

McQuarrie, D.A., 1976, *Statistical Mechanics* (Harper and Row).

Meir, Y. and Wingreen, N.S., 1992, Landauer formula for the current through an interacting electron region. In Physical Review Letters **68**, 2512.

Nikulov, A. and Sheehan D. (eds.) 2004, Entropy, **6**.

Paulsson, M. and Datta, S., 2003, Thermoelectric effect in molecular electronics. In Physical Review, **B67**, 241403(R).

Reddy, P., Jang, S.Y., Segalman, R. and Majumdar, A., 2007, Thermoelectricity in molecular junctions. In Science, **315**, 1568.

Saha, D., Holub, M., Bhattacharya, P. and Liao, Y.C., 2006, Epitaxially grown MnAs/GaAs lateral spin valves. In Applied Physics Letters, **89**, 142504.

Salahuddin, S. and Datta, S., 2006a, Electrical detection of spin excitations. In Physical Review, **B73**, 081301(R).

Salahuddin, S. and Datta, S., 2006b, Self-consistent simulation of quantum transport and magnetization dynamics in spin-torque based devices. In Applied Physics Letters, **89**, 153504.

Zhang, Qi-Ren 2007, A General information theoretical proof for the second law of thermodynamics, arXiv: quant-ph/0610005v3.

Chapter 8
Manipulating Electron Spins
in an InGaAs/InAlAs Two-Dimensional
Electron Gas

C. L. Yang, X. D. Cui, S. Q. Shen, H. T. He, Lu Ding, J. N. Wang, F. C. Zhang
and W. K. Ge*

8.1 Introduction

With the basic idea of the spin field effect transistor, as proposed by Datta and Das
(1990), the electric field tunable spin-orbit interaction has been an ideal candidate
for manipulating the electron spin-polarization. The structural inversion asymmetry (SIA) induced spin-orbit coupling (Rashba interaction) (Bychkov and Rashba,
1984) in a two-dimensional electron gas (2DEG) system has attracted more and
more attention due to its potential applications in spintronics devices. A two-
dimensional electron gas (2DEG) in a semiconductor hetero-structure with C_{2v}
symmetry and vertical structural inversion asymmetry (SIA) would induce a Rashba
spin-orbit coupling and lead to spin splitting of the conduction band in the momentum space (Bychkov and Rashba, 1984; Winkler, 2003). This system provides a
good platform for electrical control of electron spins in semiconductors. Such examples include spin photocurrent (Ganichev et al., 2002), electric-dipole-induced spin
resonance (Duckheim and Loss, 2006), and spin coherent transport (Crooker et al.,
2005; Murakami et al., 2003; Sinova et al., 2004; Shen et al., 2004). With spin
splitting of the energy bands, optical excitation of quantum well (QW) structures by
circularly polarized radiation will lead to a current whose direction and magnitude
depend on the helicity of the incident light. This is named the circular photogalvanic effect (CPGE) and has been demonstrated using inter-subband (Ganichev
et al., 2001; Ganichev and Prettl 2003) or inter-band (Bel'kov et al., 2003) excitations. A related effect, in which the current is driven by the spin-flip process of the
non-equilibrium population of electrons in the spin-split bands, is called the spin-
galvanic effect (SGE) (Ganichev et al., 2002). The most interesting consequence of
the k-dependent spin splitting is the implication that an applied electric field would
induce not only a charge current but also a spin polarization (Aronov et al., 1991;

*Department of Physics and Institute of Nano-Science and Technology, The Hong Kong
University of Science and Technology, Hong Kong, China
e-mail: phweikun@ust.hk

Z. Tang, P. Sheng (eds), *Nano Science and Technology.*
© Springer 2008

Edelstein, 1990; Chaplik et al., 2002; Inoue et al., 2003; Kato et al., 2004; Silov et al., 2004; Ganichev et al., 2004) along the direction perpendicular to the current.

This paper provides the experimental evidence of circularly polarized optical-excitation-induced spin photocurrent in (001) grown InGaAs/InAlAs 2DEGs under oblique incidence of radiation for inter-band excitation, which is 2 orders of magnitude stronger than similar observations using far-infrared excitation for inter-subband transitions. The theoretically predicted spectral inversion of the CPGE spin photocurrent was demonstrated experimentally. Furthermore, we have measured the current-induced spin-polarization in the same samples using Kerr rotation experiment. The spin splitting of the energy bands was examined by the beating of the Shubnikov–de Haas (SdH) oscillations, which reveals a unified picture for the spin photocurrent, current-induced spin polarization and spin orbit coupling.

While optically injected spin current has been evidenced both in theory and experiments (Ganichev et al., 2001, 2002; Ganichev and Prettl, 2003; Bhat et al., 2005; Li et al., 2006), but detecting spin current, which is of great interest to spintronics physics as well as device applications, remains to be a challenging problem (Hübner et al., 2003; Stevens et al., 2003; Kato et al., 2004; Sih et al., 2005; Wunderlich et al., 2005; Saitoh et al., 2006). Coherent transport and control of electron spins in semiconductor hetero-junctions have been studied intensively with the ultimate aim of implementing spintronic devices (Ganichev and Prettl, 2003; Wolf et al., 2001; Awschalom et al., 2002; Zutic et al., 2004). Among these efforts, optical injection of electron spin into 2DEG can be regarded as an efficient source to implement spin transport. In the spin photogalvanic effect, electrons in the valence band are pumped into the conduction band by irradiation of circularly polarized light, and the nonuniform distribution of the excited electrons due to the spin-orbit coupling will circulate a spin polarized electric current. In the case that the incident light is normal to the 2D plane, electrons with opposite spins would travel in opposite directions. As a result, a pure spin current circulates while the electric photo-current vanishes (Ganichev et al., 2001; Bhat et al., 2005).

In this paper, we report an observation of onward or outward electric currents in a crossbar-shaped (001) InGaAs/InAlAs quantum well when linearly polarized light irradiates on the sample at normal incidence near the junction at 77 K. Though no electric photo-current was measured when the incident light spot is far away from the crossing region of the crossbar channels, which is consistent with the conventional spin photogalvanic effect, electric currents were detected when the light spot is close to the crossing region. The electric current was demonstrated to share the same light-polarization dependence as the spin current. We attribute the observed electric current to be converted from the optically injected spin current by scattering in the crossing region.

8.2 Experimental and Results

Two samples (named D and E) studied here were $In_xGa_{1-x}As/In_{0.52}Al_{0.48}As$ 2DEGs grown on semi-insulating (001) InP substrate with well thickness of 14 nm. The SIA was achieved by δ–doping of only one side of the barrier layer (on top

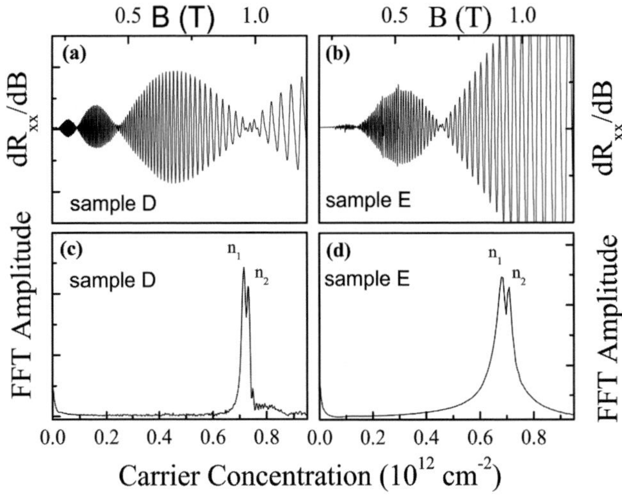

Fig. 8.1 (a) and (b) are the beating patterns of the SdH oscillations at 1.6 K under low magnetic field as shown in the first derivative of the magneto-resistance R_{xx} of samples D and E, respectively. (c) and (d) are the calculated carrier concentrations based on the FFT spectrum from (a) and (b)

of the well). To enhance the SIA, sample E was grown with a graded indium composition from 0.53 to 0.75 for the quantum well, instead of the uniform indium composition of 0.70 for sample D. Hall measurements at 1.6 K showed that the carrier concentration was $1.5 \times 10^{12}\,\mathrm{cm}^{-2}$ and $1.4 \times 10^{12}\,\mathrm{cm}^{-2}$ for samples D and E, respectively. The SdH oscillations as well as their pronounced beating patterns at low magnetic field were indicated in Fig. 8.1a and b using the first derivative of R_{xx}. The beating pattern arises from two closely separated oscillation frequency components caused by the presence of two kinds of carriers in the system. The beating leads to double-peak structures with similar height in the fast Fourier transform (FFT) spectra and gives the density of the two kinds of carriers as shown in Fig. 8.1c and d. The calculated carrier density is in excellent agreement with the Hall concentration if a spin degeneracy of 1 is taken. In our samples, the distinct beating pattern in the SdH oscillations are believed to come from the spin splitting in 2DEGs, which is further supported by the following spin photocurrent and current-induced spin polarization experiment. The obtained spin-resolved concentrations allow us to determine the Rashba spin-orbit interaction parameter α (Engels et al., 1997; Schapers et al., 1998), which gives the value of $3.0 \times 10^{-12}\,\mathrm{eVm}$ and $6.3 \times 10^{-12}\,\mathrm{eVm}$ for samples D and E, respectively, indicating that the stronger the inversion asymmetry, the larger the spin-orbit interaction in a system. The measured strongly structure-related spin-orbit coupling coefficient suggests that the SIA induced spin-splitting is the dominating mechanism.

The experimental setup for the interband-transition-induced CPGE is schematically shown in the inset of Fig. 8.2. A tunable Ti:sapphire femtosecond laser was employed for the band-to-band excitation with its polarization modified by either a crystalline $\lambda/4$ waveplate which yields right (σ^-) or left (σ^+) hand circularly

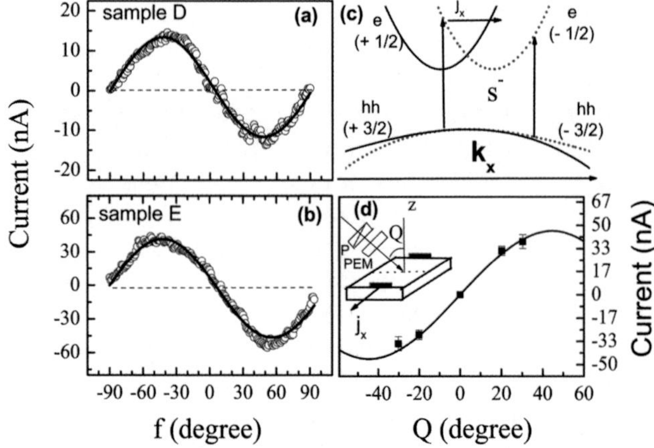

Fig. 8.2 (a) and (b) are the angle ϕ dependence of the spin photocurrent in the 2DEG sample D and E, respectively, with oblique incidence angle of 30^0 and laser power of $100\,\text{mW}$ ($\lambda = 880\,\text{nm}$) at $10\,\text{K}$, showing the helicity dependence of the photocurrent. The solid lines are the fit using $\sin 2\phi$. (c) is a schematic diagram with spin-split valence and conduction bands to show the microscopic origin of the CPGE induced photocurrent. The vertical arrows show the permitted transitions with σ^- excitation and the horizontal arrow indicates the final direction of the induced photocurrent. (d) is the incidence angle Θ dependence of the photocurrent. The solid line is the fitted curve based on Eq. (8.1). The experimental setup of the spin photocurrent is depicted in the inset

polarized light, or a PEM which yields a periodically oscillating polarization between σ^- and σ^+. The electrodes are made at the $[110]$ or $[1\bar{1}0]$ sample edges to lead the current along the laboratory x-direction. The photocurrent I_x was measured in the unbiased structures at various temperatures via a low-noise current amplifier and a lock-in amplifier. The only difference in our experimental data between those using the $\lambda/4$ waveplate and those using the PEM is that the former presents a background current even at normal incidence, while the latter does not. This is most likely due to the Damber effect when interband excitation was employed (Ganichev and Prettl, 2003). To show only the polarization dependent signal, all the data presented are measured by PEM, which gives the difference between the σ^- and σ^+ polarizations. The incident light beam is polarized in the $z - y$ plane, with an angle of Θ away from the normal. By changing the angle ϕ between the polarization plane of the incident light and the optical axis of the $\lambda/4$ waveplate or PEM, we can measure the helicity dependence of the photocurrent, which is the hallmark of spin photocurrent that distinguishes it from other photocurrent effects.

Figure 8.2a and b show the dependence of the photocurrent of the two samples on the laser beam polarization, represented by angle ϕ between the optical axis of the $\lambda/4$ modulator and the polarization plane of the incident light, with $\lambda = 880\,\text{nm}$ and $\Theta = 30^0$. It clearly shows that when ϕ changes from $-\pi/4$ to $+\pi/4$, the current varies from one maximum to the other maximum of opposite direction, in good agreement with the fitting using $\sin 2\phi$ (see the discussion below). Figure 8.2d shows the Θ (incidence angle) dependence of the spin photocurrent. When $\Theta = 0$, since there is no y-component of optically induced electron spin polarization, no current is created. With the increase of Θ, leading to a larger y-component of the electron spin polarization, the spin photocurrent first gets larger but finally gets smaller because of the increased reflection.

Ganichev and Prettl (2003) have systematically used inter-subband excitation to study the CPGE induced spin photocurrent. Inter-band excitation using circularly polarized light will also produce a spin photocurrent, for which the microscopic mechanism can be schematically described as shown in Fig. 8.2c. Considering the SIA induced Rashba interaction, there will be non-parabolic terms in the Hamiltonian, such as the linear (in $k_{//}$) spin splitting for electrons (or light holes), and spin splitting of the heavy-hole (Winkler, 2000) states proportional to $k_{//}^3$. The coupling between the spin and wave vector of the carriers, as well as the spin related selection rules, yield a non-uniform distribution of carriers in k-space upon circularly polarized optical excitation. The imbalanced momentum relaxation of electrons in the conduction band results in a net current due to the absorption of the circularly polarized light.

For quantum well structures of symmetry C_{2v} grown along the principal axis [001] with structural inversion asymmetry, a photocurrent can be generated only under oblique incidence of irradiation. If the incidence is in the (y, z) plane, then the photocurrent is induced along the x direction (Ganichev et al., 2001; Ganichev and Prettl 2003) and can be phenomenologically estimated to be

$$j_x = \gamma_{xy} t_p t_s \sin \theta \, E_0^2 P_{circ}, \tag{8.1}$$

where γ_{xy} is a pseudotensor component related to the Rashba coupling coefficient, t_p and t_s are the transmission coefficients from Fresnel's formula for linear p and s polarizations, E_0 is the electric field amplitude in vacuum, θ is the angle of refraction ($\sin \theta = \sin \Theta / \sqrt{\varepsilon^*}$, $\varepsilon^* \sim 13$). Since

$$P_{circ} = \frac{I_{\sigma^-} - I_{\sigma^+}}{I_{\sigma^-} + I_{\sigma^+}} = \sin 2\phi, \tag{8.2}$$

j_x is expected to be proportional to $\sin 2\phi$ at a fixed incidence angle Θ, just as we have observed. Considering that t_p and t_s are Θ dependent, we can also fit the Θ dependence of the experimental j_x based on Eq. (8.1) and the Fresnel's formula

$$t_p t_s = \frac{4 \cos^2 \Theta}{\left(\cos \Theta + \sqrt{\varepsilon^* - \sin^2 \Theta} \right) \left(\varepsilon^* \cos \Theta + \sqrt{\varepsilon^* - \sin^2 \Theta} \right)}. \tag{8.3}$$

The solid line in Fig. 8.2d is the fitted result, which is in excellent agreement with the experimental data. Furthermore, we find that the photocurrent is almost linearly dependent on the laser power. For sample D, the current drops faster when temperature is higher than 200 K, leaving a very small signal at 300 K. But for sample E, the current only drops about a factor of 4 from 11 K to room temperature. We believe that the larger photocurrent of sample E is due to its much larger spin splitting, as also revealed by the SdH experiment. We also find that the inter-band excitation induced spin photocurrent is up to 2 orders of magnitude stronger than that of the inter-subband transitions in similar experiments. In addition to the incident direction and the polarization dependence of the spin photocurrent, we have found that it also changes direction when the laser wavelength is changed, as indicated in Fig. 8.3 for the spectra dependence of the CPGE photocurrent. The photoreflectance spectra of the samples are shown to clearly indicate the quantized energy levels of electrons and holes as marked by the arrows. For the two samples, we have explicitly shown that the CPGE spin photocurrent will change its sign when the laser wavelength changes, which is in agreement with the theoretical prediction by Golub (2003). The spectral inversion in the CPGE spin photocurrent is thought to be a characteristic to distinguish the spin splitting mechanisms between the SIA and bulk inversion asymmetry (BIA). From the big dip in the spectra response, we can tell that the SIA should be the dominating mechanisms for the spin splitting in the samples. This is also consistent with the dramatic change of the Rashba coefficient with our modification of the sample structure as mentioned in the SdH measurement. There are mainly two contributions for spectral inversion of the spin photocurrent. The first one comes from the transitions from the light hole to the electron subband. Due to the selection rules for light hole and heavy hole (such as $-3/2 \rightarrow -1/2$ for heavy hole, $-1/2 \rightarrow +1/2$ for light hole with $\Delta j = +1$), electrons excited from the heavy hole and the light hole bands by the same circularly polarized light will jump to opposite spin splitting branches ($+1/2$ or $-1/2$) in the conduction band, which will result in reversed current in the SIA case. Secondly, as pointed out by Golub (2003), the negative contribution can also come from the heavy hole related

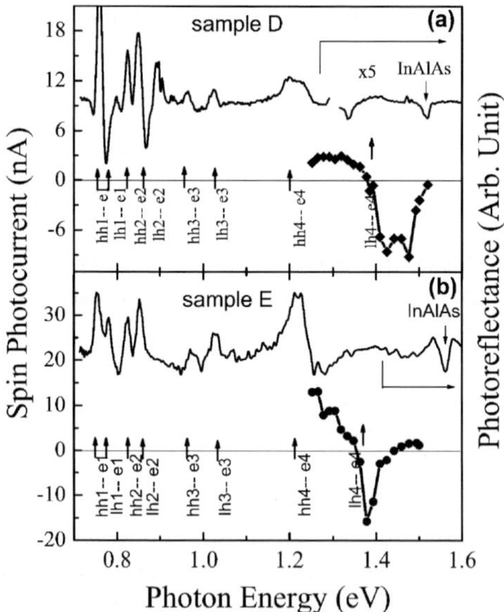

Fig. 8.3 Spectral response of the spin photocurrent for (a) sample D and (b) sample E. The photoreflectance spectra of the two samples are also shown to determine the electronic structures. The arrows indicate the heavy hole (hh) and light hole (lh) related transitions

transitions at larger wavevector due to the mixing or anti-crossing of the hole bands. The spectra inversion in the CPGE spin photocurrent is very simple and useful in distinguishing the spin splitting mechanisms between the SIA and BIA. We intend to carry out further studies along this direction, to involve ground states excitations by changing the sample composition and/or expand the laser wavelength range, and make careful calculations for comparison.

There have been theoretical predictions for spatially homogeneous spin polarization resulting from an electrical current (Edelstein, 1990) in systems such as 2DEGs, in additional to the spin Hall effect caused by the spin accumulation at the edges of a sample due to the existence of a spin current perpendicular to a charge current. Indeed we can simply interpret this current-induced spin polarization as the converse process of the spin photo-current. Since the Rashba coupling term in the Hamiltonian can be expressed as

$$H_{SO} = \alpha \vec{\sigma} \cdot \left(\vec{z} \times \vec{k}_{//} \right) = \alpha \vec{k}_{//} \cdot (\vec{\sigma} \times \vec{z}) , \qquad (8.4)$$

one can interpret this coupling as if the momentum (carried by a current) induces an effective magnetic field, or the spin induces an effective momentum. Inoue et al. (2003) first derived the diffusive conductance tensor for a disordered 2DEG with spin-orbit interaction and showed that the applied bias (E_x) induces a spin accumulation

$$\langle S_y \rangle = 4\pi e \tau D \lambda E_x, \qquad (8.5)$$

where $D = m_e / 2\pi \hbar^2$ is the 2DEG density of states per spin, the lifetime τ is the momentum relaxation time, and $\lambda = \alpha \langle E_Z \rangle / \hbar$ represents the Rashba interaction. From Eq. 8.3 we find that the intensity of the current-induced spin polarization will also give a measure of the spin-orbit interaction parameter α. Kerr effect was used to monitor the spin polarization along the y-direction under an external electric field in the x-direction, as shown in the inset of Fig. 8.4b. In the Kerr rotation experiment, a polarizer and an analyzer were placed in the incident and reflected laser beams, respectively. A PEM situated before the analyzer on the reflected beam was employed to sensitively monitor any optical polarization rotation induced by the spin polarization in the samples. To get rid of any possible reflectivity changes induced by heating effects due to the applied electric field, the Kerr rotation was extracted from the signal difference between positive and negative electric fields. The laser wavelength was tuned to be close to the transition energy between the third excited state of heavy hole and the electron, which was identified from photoreflectance measurements. Figure 8.4a and b show the Kerr rotation against the external electric field at 11 K for samples D and E, respectively. The detected Kerr signal increases with increase of the external field, which is in agreement with the theoretical prediction that the spin polarization S_y is proportional to E_x. However, we didn't observe linear relationship between the spin polarization and external field, which can originate from two possibilities. Firstly, nonlinearity at low field mostly comes from the experimental uncertainty due to the limited sensitivity.

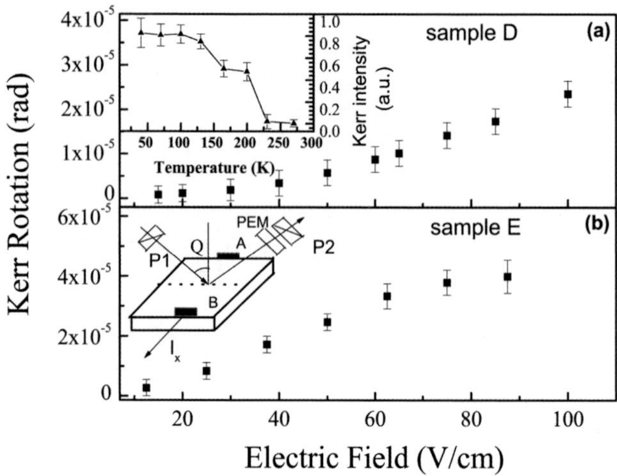

Fig. 8.4 Electric field (E_x) dependence of the Kerr rotation of sample D (a) and E (b) to show the current induced spin polarization $\langle S_y \rangle$ at 10 K. The temperature dependence of the Kerr rotation of sample D ($E_x = 50$ V/cm) is shown in the inset of (a). The experimental setup is schematically shown in the inset of (b), of which P1 and P2 represent the polarizer and analyzer, respectively

The second reason is the electric field dependent spin relaxation time. The spin relaxation time is found to decrease (Kato et al., 2004) with increase of the electric field, resulting in a reduced spin polarization or spin density, which is the reason that Kerr rotation signal gets saturated at high field. Also we find that sample E gives a Kerr rotation about 3–4 times larger than that of sample D, which is consistent with the much larger spin-orbit coupling coefficient in the former as determined by the SdH oscillation. The temperature dependence of the Kerr rotation, as presented in the inset of Fig. 8.4a for sample D, shows that the spin polarization drops quickly when temperature is above 100 K, which is due to the reduced spin and momentum relaxation time at elevated temperatures. The experimental setup for converting spin current into electric current is shown in Fig. 8.5b, with a crossing bar carved on the above sample E. The Rashba coupling coefficient is about $\alpha = 3.0 \times 10^{-12}$ eVm determined by Shubnikov-de Hass oscillation (Golub, 2003). The time-resolved Kerr rotation measurement with standard pump probe technique gives the spin decoherence time $\tau_s \approx 106$ ps at 77 K as shown in Fig. 8.5a. The Fermi velocity of the sample is estimated to be $v_F \approx 5.0 \times 10^5$ m/s (taking the effective mass as $0.07\,m_e$). Thus the spin coherence length is roughly estimated to be $l_s \approx 70$ μm at 77 K. Taking consideration of these parameters we designed a setup with two electric channels of widths WS $= 20$ μm and WL $= 200$ μm, respectively, one being narrower and the other wider than the spin coherence length, i.e., $WS < l_s < WL$, as shown in Fig. 8.5b. The planar electric channels were carved out by standard photo-lithography and wet etching along [110] and [1⁻10], defined as lab x- and y-axis respectively (see Fig. 8.5b). Electric contacts were made to the 2DEG by Ni/Au metal electrodes at far ends of the electric channels.

Linearly polarized laser light of wavelength 880 nm and power 40 mW produced by a tunable Ti:sapphire laser was used as the exciting source for the inter-band

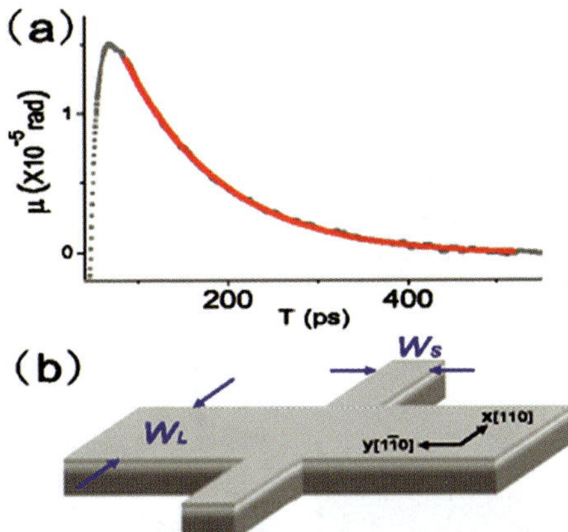

Fig. 8.5 (a) The data of time resolved Kerr rotation for spin decoherence was fitted by an exponential function $\exp[-t/\tau_s]$ (the red line) with the spin decoherence time $\tau_s = 106$ ps. (b) Geometry of the asymmetric crossbar-shaped 2DEG sample, where the spin coherence length is expected to lie between the width of the narrow and wide channels

transition. The incident light was firstly guided through a photo-elastic modulator (PEM) which modulated the light polarization to be oscillating between two orthogonal directions with a frequency of 50 KHz. Then the light was focused through a $10\times$ objective lens before being normally shed upon a $10\,\mu$m diameter spot on the sample which was mounted in a cryostat at liquid nitrogen temperature. The electric currents passing through the terminals were monitored by voltage drops at symmetrically loaded resistors of $633\,\Omega$. The voltages were read out with a lock-in amplifier which was triggered by the PEM so that the electric currents induced by non-polarization related phenomena e.g. Dember effect and thermoelectric effects were clearly eliminated.

No measurable current along the wide channels (y-axis) was observed on the background noise of tens of picoamperes under normal incidence, though a significant photocurrent was observed if the circular polarized light was shed at an oblique angle on the same samples (Yang et al., 2006). However, when light with linear polarization along the x- or y-axis was normally incident onto the plane near the electric channel junction, we observed reproducible currents along the narrow channels under unbiased conditions. Figure 8.6c–e show the currents passing through the terminals as functions of the spot position when the incident light is along the $20\,\mu$m-wide channel (x-axis). The current peaks at $x = 0$ and $200\,\mu$m corresponding to the positions of the two channel junctions, and vanishes at a distance of $50\,\mu$m away from the peak position. Note that the current curves of I_{C-A} and I_{C-B} and of I_{A-D} and I_{D-B} almost overlap, indicating that the current flows inward or outward simultaneously through the channel junction. The spatial distribution of the observed current is substantially different from the current due to Hall effect or the reciprocal spin Hall effect, where the Hall current is unidirectional. The flow pattern of the currents and the absence of a net electric current along the wide channel are the signatures of the observed effect as summarized in Fig. 8.6b. The vanishing of

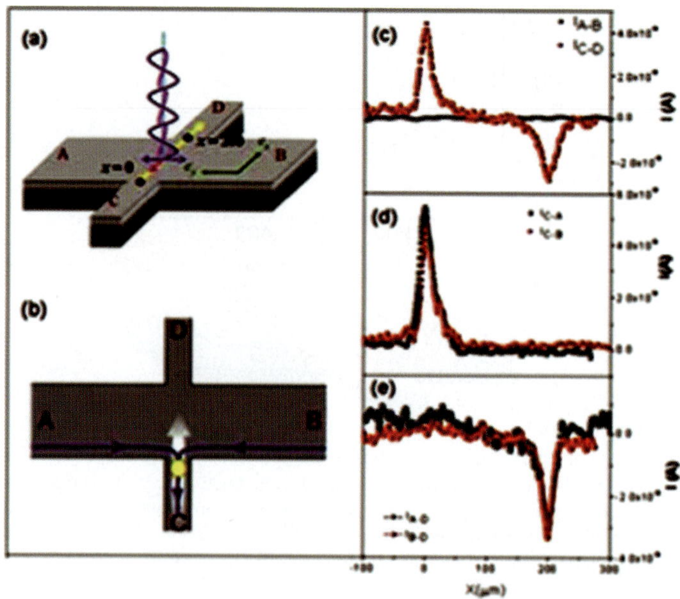

Fig. 8.6 (a) Schematic view of the 2DEG sample and the experimental setup. The 20 and 200 μm-wide planar channels are carved out by standard photolithography and wet etching. A linearly polarized light beam, polarized along x-axis as indicated by red color or along y-axis as indicated by blue color, scans along the narrow channel (x-axis) at normal incidence. (b) Flow pattern of measured electric currents for light spot along the narrow channel. (c), (d), and (e) show the typical charge currents through all the terminals as a function of the light spot position at x-axis

all currents at the middle point of the C−D channel, $x = 100\,\mu$m, evidently indicates that the observed phenomena are not due to conventional photocurrents. Note that the spatial location of the light spot depicted in Fig. 8.6 is hundreds of microns away from the metal electrodes, therefore any contribution to the measured electric currents from the interfaces between the metal electrodes and semiconductor channels is negligible. As the current measurements were locked at the polarization oscillating frequency, the obtained current measures the difference between the electric currents generated by the two orthogonally polarized lights in the device. To further examine the relationship between the observed currents and polarization of the incident light, we rotate the polarization plane of the incident light while the light spot is fixed at the peak position of I_{C-A} and I_{C-B} in Fig. 8.6c. The current varies with Φ, the angle between the light polarization plane and the y-axis, and a good fit is obtained with the form $\cos 2\Phi$, as shown in Fig. 8.7. Obviously the polarization dependence of the current is in excellent agreement with the calculated angle dependence of the spin current (Li et al., 2006; Zhou and Shen, 2007). Note that the lock-in technique we used does not measure any Φ-independent current because of cancellation.

The result of the laser scan along the edge of channel A−B at $x = 10\,\mu$m is shown in Fig. 8.4. Electric currents I_{A-C} and I_{B-C} were observed near the channel junction, and they were almost identical. The current I_{A-B} remains negligible

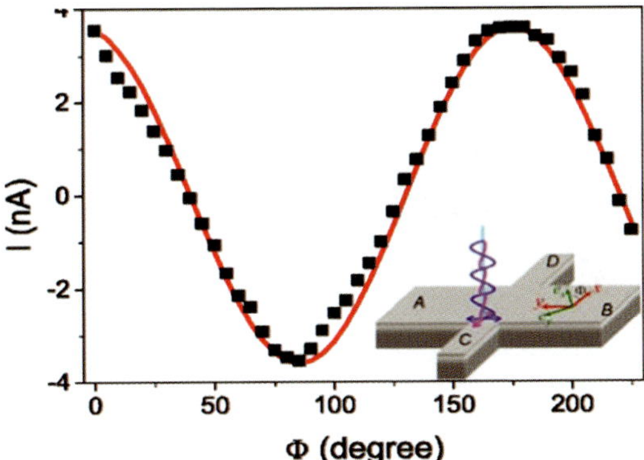

Fig. 8.7 The charge current I_{A-C} (black dot with error bars) at channel junction $x = 0$ varies with the polarization angel $\sim$ of light relative to the lab frame, as shown in the inset. The red curve is a fit of $\cos 2\Phi$

despite the laser spot position moving. The spatial dependence of the current are drastic in the vicinity of $Y = 0$, or along the central line of channel C−D. When the laser scans along the central line of channel A−B, i.e., at $x = 100\,\mu m$, no current was observed, for the light spot is located well beyond the spin coherence distance. This is consistent with the conventional spin photogalvanic effect, where the current vanishes when the linearly polarized light falls normally on the plane. We thus conclude that the observed electric current near the junction with a narrow width ($20\,\mu m$) is not a conventional photocurrent (Ganichev et al., 2001, 2002; Yang et al., 2006). The equality of I_{A-C} and I_{B-C} obviously excludes the possibility of a diffusion mechanism for generating the photo-current.

To eliminate the possible effect of the specific sample, we also carried out measurements on sample D, with a stronger Rashba coupling about $\alpha = 6.3 \times 10^{-12}\,eVm$. Essentially the same results were observed. To understand the experimental results, we interpret the observed electric current as a result of the scattering of the spin current near the junction of the crossing bars. Firstly, the optical excitation serves as the source of a spin current with its spin polarization lying in the plane and perpendicular to the direction of the electron motion. In detail, the linearly polarized light selectively pumps electrons in the valence band up to the conduction band which is split due to the strong spin-orbit coupling. As a result the excited electrons with non-zero spin-dependent momenta form a spin current before recombining with holes in the valence band. The effective distance of this spin current is estimated to be within the spin coherence length l_s, which has the relation of WL/2 > l_s > WS/2. Keeping this picture in mind the observed effect can then be well understood as follows. In the case that the light spot scans along the central line of the narrow channel C−D in Fig. 8.6, the geometric boundaries of the sample set a limit to the motion of the excited electrons and lead them to two opposite directions with opposite in-plane spin polarizations. Two beams of the excited electrons compose the same spin current, which is then scattered by the edges of the crossing region to produce a transverse electric current. On the other hand, when the light spot scans along the central line of the wide channel A−B in

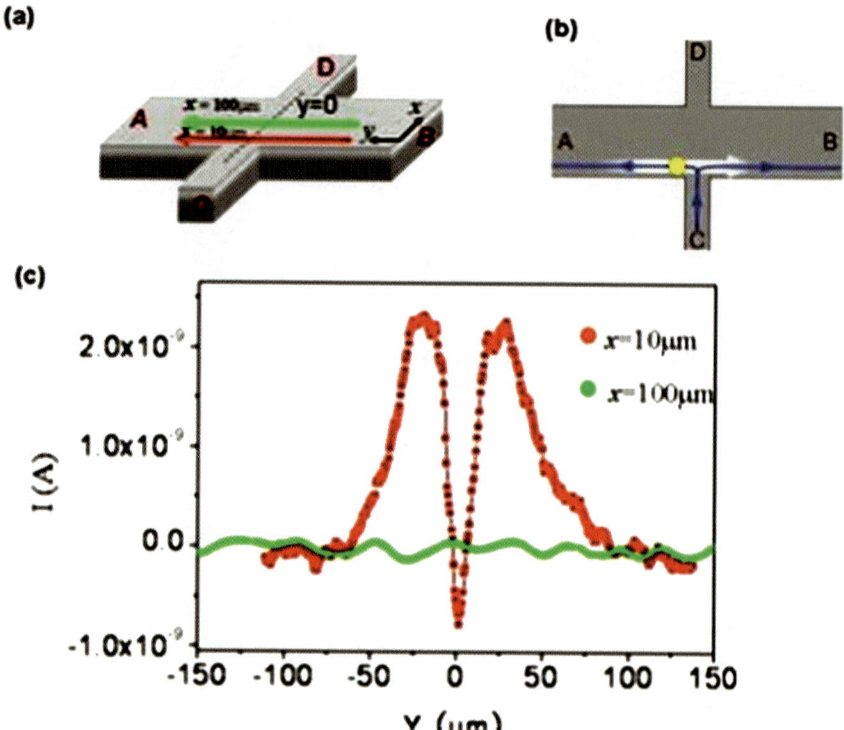

Fig. 8.8 (a) Schematic view of the light scanning along the 200 μm-wide channel. The red and green scanning lines are parallel to y-axis and are 10 and 100 μm away from the channel edge respectively. (b) Flow pattern of measured electric current for light spot scanning along the edge. (c) The corresponding charge currents I_{A-C} show strong position dependence near the junction center

Fig. 8.8, the spin current decays to zero before it reaches the edges of the crossing region, therefore no electric current circulates. While in the case that the light spot moves along the edge, also shown in Fig. 8.8, when the distance between the spot and the edge is within the distance of l_s, the edge may again impose a constraint to the movement of the electrons and lead to an outward (inward) electric current along the edge, together with an inward (outward) electric current perpendicular to the edge, and thus a measurable current between $C-A/C-B$ (in the case the edge is close to terminal C) or $D-A/D-B$ (in the case that the edge is close to D). As it is known from the optical Bloch equation (Zhou et al., 2007). that the spin current depends on the angle Φ between the edge and the polarization direction of light, $J_{xy} = J_0 + J_1 \cos 2\Phi$, the same dependencies on the light polarization of the electric current as shown in Fig. 8.7 can be considered as a direct evidence to strongly support the above scenario. Furthermore, the sample is carved on a uniform strip of InGaAs/InAlAs, so that the spin injection problem is clearly avoided when the spin current passes through the scattering region at the junction. Also as the spin coherence length at liquid nitrogen temperature is longer than the width WS = 20 μm of the narrow channel, the impurity effect or other interactions would not be crucial in this experiment. Finally, as the phase coherence length of electrons in this system is much shorter than WS, the possibility that this effect being a mesoscopic one should be excluded.

8.3 Discussion and Summary

In summary, we have shown two converse effects, spin photocurrent and current-induced spin polarization, in a specially designed Rashba system, i.e. a 2DEG system with asymmetric potential barriers. It is worth noticing the comparison of the two samples used in this work. The Rashba coefficient measured from SdH oscillations has a ratio of about 2 between samples E and D. For the spin photocurrent, the efficiency ratio is a bit less than 3 between these two samples. For the Kerr rotation, the relative efficiency ratio is about 3 to 4. This evidently shows that the physical mechanisms behind the three effects are uniformly based on the Rashba coupling, and any of them can be taken as an indication of the Rashba effect. Though the general trends are consistent, we do not expect to find precise values for these ratios, as so many parameters are involved in any quantitative comparison. In particular, as shown in Eq. 8.5, the momentum relaxation time is involved in evaluating the current-induced spin polarization, so that more work in evaluating the lifetime will be helpful to further elaborate the correlation between various spin-splitting induced effects. As the CPGE was observed under obliquely incident excitation of inter-band transitions, the spin photocurrent was found to have efficiency much larger than the similar effect induced by inter-subband transitions. Meanwhile, the Kerr effect induced by a current flow in the same samples is quite pronounced as well, indicating the possibility of driving spin polarization with a current. The current-induced spin polarization based on the Rashba effect may be a useful spin source for spintronic devices, in addition to magnetization under magnetic field. Finally, the spectral inversion in the CPGE spin photocurrent provides us an easy way to distinguish the SIA and the BIA in the spin-orbit coupling.

The conversion of spin current to electric current has been discussed extensively in the context of the reciprocal spin Hall effect where the seperated spins are out of plane (Hirsch et al., 1999). In the present case the spin current is polarized within the plane, as the linearly polarized light does not generate out-of-plane spin polarization due to conservation of the total angular momentum. As examined in (Li et al., 2006), an electron of spin-y polarization will be equally scattered along the x and $-x$ axes, and hence the resulted scattered electric current has such a pattern: either both outward or both inward, in the transverse directions of the spin current. Our experimental observation of the electric currents is in excellent qualitative agreement with this picture. In the Landauer-Büttiker formalism, numerical simulations were performed in a symmetric crossbar shaped setup with the tight binding approximation. The numerical results and symmetry analysis show that the resulted electric current depends strongly upon the spin polarization of the spin current. For the case of the in-plane-polarized spin current as in the present experiment, the conversion rate for the spin current to electric current is estimated to be $0.3 \sim 1\%$ (Li et al., 2006). In the present work, although there are some unknown parameters of the sample, the amplitude of the injected spin current J_{xy} by the linearly polarized light can be roughly estimated by using the measurable photocurrent J_x injected by the circularly polarized light at a small oblique angle Θ away from normal. If the powers of the applied two laser lights are equal, a theoretical estimation gives $J_x(\Theta)/J_{xy} \approx 0.17\Theta(2e/\hbar)$ by solving of semiconductor Bloch optical equation near the Γ point of the InGaAs 2DEG (Zhou and Shen, 2007). Taking the light spot size, the geometry of the setup, and the laser power into account, we estimate from the experimental data that $J_x(\Theta) \approx 94\Theta$ nA (Yang et al., 2006), hence the injected spin current is estimated to be approximately 550 nA (h/2e). The observed electric current in the present experiment is several nA, say 4.0 nA for the peaks in Fig. 8.6c. The conversion rate from the spin current to charge current is roughly about 0.8%, close to the numerically simulated values.

Briefly speaking, we observed onward or outward electric currents in a crossbar-shaped (001) InGaAs/InAlAs quantum well when linearly polarized light irradiates on the sample at normal incidence near the junction. Both the spatial pattern and polarization dependence of the electric current are in agreement with the scenario that the electric current is produced by the optically injected spin current scattered near the crossing edges. Our observation provides a realistic technique to detect spin currents, and opens a new route to study spin-related science and engineering in semiconductors.

Acknowledgments The authors thank S. J. Xu, L. Ding and X. Z. Ruan for assistance in laboratory facility and X. Dai, J. Wang, B. Zhou, S. C. Zhang and Q. Niu for helpful discussions. This work was supported by the Research Grant Council of Hong Kong under Grant No.: HKU 7039/05P, the Large-Item-Equipment Funding and the University Development Funding of HKU on "Nanotechnology Research Institute" program under contract No. 00600009. The university grant DAG 04/05.SC 21 and 05/06.SC30 from HKUST. After completion of the present work we were beware of relevant work by Valenzuela and Tinkham (2006), where the electric current induced by spin current of the out-of-plane polarization in metal Al was detected experimentally. We also greatly appreciate Prof. B. A. Foreman for his help and his critical reading of the manuscript.

References

Aronov, A.G., Lyanda-Geller, Y.B., Pikus, G.E., 1991, *Soviet Physics - JETP*, **73**, pp. 537–541.

Awschalom, D.D., Loss, D., Samarth, N. (eds) 2002, *Semiconductor Spintronics and Quantum Computation* (Springer, Berlin).

Bel'kov, V.V., Ganichev, S.D., Schneider, P., Back, C., Oestreich, M., Rudolph, J., Hägele, D., Golub, L.E., Wegscheider, W., Prettl, W., 2003, Circular photogalvanic effect at inter-band excitation in semiconductor quantum wells. *Solid State Communications*, **128**, pp.283–286

Bhat, R.D.R., Nastos, F., Najmaie, A., Sipe, J.E., 2005, Pure spin current from one-photon absorption of linearly polarized light in noncentrosymmetric semiconductors. *Physical Review Letters*, **94**, p. 096603.

Bychkov, Y.A., Rashba, E. I., 1984, Oscillatory effects and the magnetic susceptibility of carriers in inversion layers. *Journal of physics C: Solid state physics*, **17**, pp. 6039–6045.

Chaplik, V., Entin, M.V., Magarill, L.I., 2002, Spin orientation of electrons by lateral electric field in 2D system without inversion symmetry. *Physica E: Low-dimensional Systems and Nanostructures*, **13**, pp. 744–747.

Crooker, S.A., Furis, M., Lou, X., Adelmann, C., Smith, D.L., Palmstrom, C.J., Crowell, P.A., 2005, Imaging spin transport in lateral ferromagnet/semiconductor structures. *Science*, **309**, pp. 2191–2195.

Datta, S., Das, B., 1990, Electronic analog of the electro-optic modulator. *Applied Physics Letter*, **56**, pp. 665–667.

Duckheim, M., Loss, D., 2006, Electric-dipole-induced spin resonance in disordered semiconductors. *Nature Physics*, **2**, pp. 195–199.

Edelstein, V.M., 1990, Spin polarization of conduction electrons induced by electric current in two-dimensional asymmetric electron systems. *Solid State Communications*, **73**, pp.233–235.

Engels, G., Lange, J., Schapers, Th., Luth, H., 1997, Experimental and theoretical approach to spin splitting in modulation-doped $In_xGa_{1-x}As$/InP quantum wells for $B \to 0$. *Physical Review B*, **55**, pp. R1958–R1961.

Ganichev, S.D., Prettl, W., 2003, Spin photocurrents in quantum wells. *Journal of physics C: Solid state physics*, **15**, pp. R935–R983.

Ganichev, S.D., Danilov, S.N., Eroms, J., Wegscheider, W., Weiss, D., Prettl, W., Ivchenko, E.L., 2001, Conversion of spin into directed electric current in quantum wells. *Physical Review Letters*, **86**, pp. 4358–4361.

Ganichev, S.D., Danilov, S.N., Schneider, P., Bel'kov, V.V., Golub, L.E., Wegscheider, W., Weiss, D., Prettl, W., 2004, Can an electric current orient spins in quantum wells? arXiv:cond-mat/0403641v1.

Ganichev, S. D., Ivchenko, E.L., Bel'kov, V.V., Tarasenko, S.A., Sollinger, M., Weiss, D., Wegscheider, W., Prettl, W., 2002, Spin-galvanic effect. *Nature*, **417**, pp. 153–156.

Golub, L.E., 2003, Spin-splitting-induced photogalvanic effect in quantum wells. *Physical Review B*, **67**, p. 235320.

Hirsch, J.E., 1999, Spin Hall Effect. *Physical Review Letters*, **83**, pp. 1834–1837.

Hübner, J., Rühle, W.W., Klude, M., Hommel, D., Bhat, R.D.R., Sipe, J.E., Driel, H.M.van, 2003, Direct observation of optically injected spin-polarized currents in semiconductors. *Physical Review Letters*, **90**, pp. 216601.

Inoue, J., Bauer, G.E.W., Molenkamp, L.W., 2003, Diffuse transport and spin accumulation in a Rashba two-dimensional electron gas. *Physical Review B*, **67**, p. 033104.

Kato, Y.K., Myers, R.C., Gossard, A.C., Awschalom, D.D., 2004a, Current-induced spin polarization in strained semiconductors. *Physical Review Letters*, **93**, p. 176601.

Kato, Y.K., Myers, R.C., Gossard, A.C., Awschalom, D.D., 2004b, Observation of the spin Hall effect in semiconductors. *Science*, **306**, pp. 1910–1913.

Li, J., Dai, X., Shen, S.Q., Zhang, F.C., 2006, Transverse electric current induced by optically injected spin current in a cross-shaped InGaAs/InAlAs system. *Applied Physics Letter*, **88**, p. 162105.

Murakami, S., Nagaosa, N., Zhang, S.C., 2003 Dissipationless quantum spin current at room temperature. *Science*, **301**, pp. 1348–1351.

Saitoh, E., Ueda, M., Miyajima, H., Tatara, G., 2006, Conversion of spin current into charge current at room temperature: Inverse spin-Hall effect. *Applied Physics Letter*, **88**, p. 182509.

Schapers, Th., Engels, G., Lange, J., Klocke, Th., Hollfelder, M., Luth, H., 1998, Effect of the heterointerface on the spin splitting in modulation doped $In_xGa_{1-x}As$/InP quantum wells for $B \to 0$. *Journal of Applied Physics*, **83**, pp. 4324–4333.

Shen, S.Q., Ma, M., Xie, X.C., Zhang, F.C., 2004, Resonant spin Hall conductance in two-dimensional electron systems with a Rashba interaction in a perpendicular magnetic field. *Physical Review Letters*, **92**, p. 256603.

Sih, V., Myers, R.C., Kato, Y.K., Lau, W.H., Gossard, A.C., Awschalom, D.D., 2005, Spatial imaging of the spin Hall effect and current-induced polarization in two-dimensional electron gases. *Nature Physics*, **1**, pp. 31–35.

Silov, A. Yu., Blajnov, P.A., Wolter, J.H., Hey, R., Ploog, K.H., Averkiev, N.S., 2004, Current-induced spin polarization at a single heterojunction. *Applied Physics Letter*, **85**, pp. 5929–5931.

Sinova, J., Culcer, D., Niu, Q., Sinitsyn, N.A., Jungwirth, T., MacDonald, A.H., 2004, Universal intrinsic spin Hall effect. *Physical Review Letters*, **92**, p. 126603.

Stevens, M.J., Smirl, A.L., Bhat, R.D.R., Najmaie, A., Sipe, J.E., Driel, H.M.van, 2003, Quantum interference control of ballistic pure spin currents in semiconductors. *Physical Review Letters*, **90**, p. 136603.

Valenzuela, S.O., Tinkham, M., 2006, Direct electronic measurement of the spin Hall effect. *Nature*, **442**, pp. 176–179.

Winkler, R., 2000, Rashba spin splitting in two-dimensional electron and hole systems. *Physical Review B*, **62**, pp. 4245–4248.

Winkler, R., 2003, *Spin-Orbit Coupling Effects in Two-Dimensional Electron and Hole Systems* (Springer, Berlin).

Wolf, S.A., Awschalom, D.D., Buhrman, R.A., Daughton, J.M., Molnár, S. von, Roukes, M.L., Chtchelkanova, A.Y., Treger, D.M., 2001, Spintronics: A spin-based electronics vision for the future. *Science*, **294**, pp. 1488–1495.

Wunderlich, J., Kaestner, B., Sinova, J., Jungwirth, T., 2005, Experimental observation of the spin-Hall effect in a two-dimensional spin-orbit coupled semiconductor system. *Physical Review Letters*, **94**, p. 047204.

Yang, C.L., He, H.T., Ding, L., Cui, L.J., Zeng, Y.P., Wang, J.N., Ge, W.K., 2006, Spectral dependence of spin photocurrent and current-induced spin polarization in an InGaAs/InAlAs two-dimensional electron gas. *Physical Review Letters*, **96**, p. 186605.

Zhou, B., Shen, S.Q., 2007, Deduction of pure spin current from the linear and circular spin photogalvanic effect in semiconductor quantum wells. *Physical Review B*, **75**, p. 045339.

Zutic, I., Fabian, J., Sarma, S.D., 2004, Spintronics: Fundamentals and applications, *Reviews of Modern physics*. **76**, pp. 323–410.

Chapter 9
Continuum Modelling of Nanoscale Hydrodynamics

Ping Sheng*, Tiezheng Qian and Xiaoping Wang

9.1 Introduction

It may come as a surprise to most physicists and material scientists that up until recently, continuum hydrodynamics can not accurate model fluid(s) flow at the nanoscale. The problem lies not in the Navier Stokes (NS) equation, which expresses momentum balance and therefore must be valid, but in the boundary condition at the fluid-solid interface, required for the solution of the NS equation. Traditionally the no-slip boundary condition (NSBC) has always been used to solve hydrodynamic problems, which states that there can be no relative motion at the fluid-solid interface. The purposes of this article are to delineate the problem and to present its resolution through the application of Onsager's principle of minimum energy dissipation, which underlies almost all the linear response phenomena in dissipative systems.

Even though lacking in first principles support, NSBC was widely regarded as a cornerstone in continuum hydrodynamics, owing to its broad applicability in diverse fluid-flow problems. However, an important exception was pointed out some time ago in the so-called the moving contact line (MCL) problem in immiscible flows. Here the contact line is defined as the intersection of the two-phase immiscible fluid-fluid interface with the solid wall. In 1974, Dussan and Davis showed with rigor that under the assumptions of (1) fluid incompressibility, (2) rigid and flat solid wall, (3) impenetrable fluid-fluid interface, and (4) NSBC, there is a velocity discontinuity at the MCL, and the tangential force exerted by the fluids on the solid wall in the vicinity of the MCL is infinite. Subsequently, by employing molecular dynamic (MD) simulations, Koplik et al. (1988) and Thompson and Robbins (1988) have shown that near-complete slip occurs at the MCL, in direct contradiction to the NSBC. The failure of the NSBC and the lack of a viable alternative mean that the usual continuum hydrodynamics can not accurately model fluids flow on the micro/nanoscale. This has often been referred to as an example of the so-called "breakdown of continuum."

---

* Department of Physics and Institute of Nano Science & Technology, Hong Kong University of Science and Technology, Clear Water Bay, Kowloon, Hong Kong, China.
e-mail: sheng@ust.hk

Z. Tang, P. Sheng (eds), *Nano Science and Technology.*
© Springer 2008

Faced with this challenge to the NSBC, there are two choices. The first is to retain the NSBC and regard the slip at the MCL to arise from the "fracturing" of the fluid-solid interface. In this scenario, the very large tangential stress in the vicinity of the MCL can break the fluid-solid bonds that are responsible for the NSBC, and the resulting slip hence lies outside the realm of hydrodynamics, just as solid fractures can not be described by linear elasticity. Thus the slip occurs only in the immediate molecular vicinity of the MCL, while NSBC holds elsewhere. The problem with this scenario is that since the conclusion of Dussan and Davis is independent of the fluid velocity, at very low velocities the tangential stress can be very small. And if the molecular dimension offers a lower cutoff to the divergence in force, that force may not be large enough to break any molecular bond, and the scenario becomes inconsistent with itself.

The other possible scenario is that hydrodynamics holds everywhere, including the MCL. But in that case the obvious question is: What is the relevant boundary condition that would resolve the force divergence and yield the near-complete slip at the MCL? Of the two choices, this second scenario is generally regarded as more daunting, since success would mean that the NSBC is at best only an approximation. In other words, success in finding the alternative boundary condition(s) is tantamount to a direct challenge to the validity of the NSBC.

We offer a resolution to the MCL problem as posed by Dussan and Davis, along the line of the second scenario (Qian et al., 2003, 2004, 2006). What differentiates the new boundary conditions from the NSBC is their derivation, in conjunction with the Stokes equation (from which the Navier-Stokes equation can be obtained by including inertial forces), from the basic principle that underlies the linear response theories in dissipative systems—Onsager's principle of minimum energy dissipation (Onsager, 1931a,b). The viewpoint expounded by this derivation is that fluid and solid are composed of molecules, albeit with different interactions. Hence if there is viscous dissipation in bulk fluid, there should be a similar *form* of dissipation at the fluid-solid interface, characterized by some parameter that would play the role of viscosity in the bulk. By doing so the present approach derives both the equations of motion and the boundary conditions within a unified framework. The relevance to the nanoscale is that the liquid-solid interfacial dissipation is characterized by a coefficient β with the dimension of [viscosity]/[length]. In particular, if one divides the bulk viscosity by β, the resulting length scale, denoted the slip length ℓ_s, is generally in the nanoscale. Hence the ratio ℓ_s/L, where L is some length scale intrinsic to the system, e.g., solid particle size in colloids, determines whether the NSBC is a good approximation (when $\ell_s/L \ll 1$), or the new boundary conditions are needed (when $\ell_s/L \sim 1$).

In what follows, a brief recapitulation of the MCL history in Section 9.2 is followed by the statement of results and comparison with MD simulations in Section 9.3. The derivation of the relevant boundary conditions will be preceded by a short introduction to the principle of minimum energy dissipation in Section 9.4. In Section 9.5 the continuum hydrodynamics formulation is shown to emerge variationally as the Euler-Lagrange equations from the dissipative energy functional. In Section 9.6 we compare the numerical predictions of our continuum hydrodynamics with MD simulation results, and discuss the implications.

9.2 History of the MCL Problem

Since the MCL represents our incision point on the question of hydrodynamic boundary condition, it is only appropriate to acknowledge the many prior and contemporary works on this problem. As shown in Fig. 9.1, when one fluid displaces another immiscible fluid, the contact line moves relative to the solid wall. The incompatibility of the NSBC with the MCL has been pointed out by Moffat (1964), Huh and Scriven (1971), Dussan and Davis (1974), Dussan (1976), and de Gennes (1985). But it was Dussan and Davis (1974) who showed that in the continuum formulation of the problem with the usual assumptions, the tangential viscous stress would vary as $\eta V/x$ along the wall, where η is the viscosity, V the wall speed, and x is the distance along the wall away from the MCL. This variation directly leads to a non-integrable singularity, implying infinite dissipation. There have been many models aimed at resolving the problem. These include the kinetic adsorption/desorption model by Blake and Haynes (1969), the various slip models of Hocking (1977), Huh and Mason (1977), and Zhou and Sheng (1990), the interface formation/disappearance model of Shikhmurzaev (1997), the diffuse-interface models by Seppecher (1996), Jacqmin (2000), Chen et al. (2000), Pismen and Pomeau (2000), and Briant and Yeomans (2004), plus the variational model of Glasner (2005). It should be noted that none of these models can reproduce the so-called slip profile (see Fig. 9.2) obtained from MD simulations.

In a separate approach, Cox divided the MCL problem into an inner region and an outer region (Cox, 1986). By carrying out an asymptotic analysis, he found to the leading order of the capillary number $Ca = \eta V/\gamma$ a relation between the apparent contact angle (the angle of the fluid-fluid interface at some mesoscopic distance away from the solid boundary) in the outer region and the microscopic contact angle, the capillary number, and the distance from the contact line. What happens in the

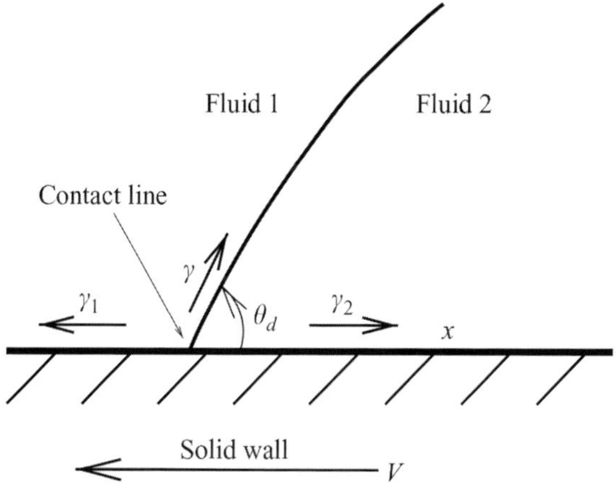

Fig. 9.1 When one fluid displaces another immiscible fluid, the contact line is moving relative to the solid wall. Owing to the contact-line motion, the dynamic contact angle θ_d deviates from the static contact angle θ_s which is determined by the Young equation $\gamma \cos \theta_s + \gamma_2 = \gamma_1$

is the quantity to be minimized if we want to maximize the probability of transition with respect to α'. For small Δt, it is seen that instead of minimizing A with respect to the target state α', the same is achieved my minimizing with respect to the *rate* $\dot{\alpha}$. Indeed, if we carry out the simple minimization on the right hand side of Eq. (9.12), we obtain the force balance equation

$$\eta\dot{\alpha} = -\frac{\partial F(\alpha)}{\partial \alpha},\tag{9.13}$$

i.e., the Langevin equation without the stochastic force term. This is reasonable, since the stochastic force has a zero mean, so Eq. (9.13) is true on the average.

Thus we learn from the above that (a) there can be a variational functional, of which the quantity A is the one-variable version, which should be minimized with respect to the rates, (b) the result of the such minimization would guarantee the force balance on average, and (c) the minimization would also yield the equations of motion and the related boundary conditions, which represent the *most probable course of a dissipative process*. The last statement essentially guarantees that in the statistical sense, the most probable course will be the only course of action observed macroscopically.

For the general case of multivaribles, the so-called dissipative variational functional can be simply generalized from Eq. (9.12) as

$$A = \frac{1}{2}\sum_{i,j}\eta_{ij}\dot{\alpha}_i\dot{\alpha}_j + \sum_{i=1}^{n}\frac{\partial F(\alpha_1,\cdots,\alpha_n)}{\partial\alpha_i}\dot{\alpha}_i,\tag{9.14}$$

where in the case of $\alpha_i's$ being field variables, the summation should be replaced by integrals, and partial derivatives by functional derivatives. In Eq. (9.14) the dissipation coefficient matrix elements η_{ij} must be symmetric with respect to the interchange of the two indices, as shown by Onsager (1931a,b) based on microscopic reversibility.

9.5 Variational Derivation of the Fluid-Solid Boundary Conditions

Armed with the variational principle, in this section we intend to first construct the energy dissipation functional A, from which desired boundary conditions and the equations of motion can be derived. As seen from Eq. (9.14), there are two parts to the functional. For the first part, we have to identify the various sources of dissipation. For the second part, we have already written down the relevant free energies in Section 9.3, what is required is to write down their variation with time.

There are basically four dissipation sources: the bulk viscous dissipation, the dissipation due to inter-diffusion at the fluid-fluid interface, the frictional dissipation at the fluid-solid interface, and the relaxational dissipation around the contact line, also due to inter-diffusion (Qian et al., 2006). For the viscous dissipation, it is well known that the total dissipation rate may be written as

$$R_{vis} = \int d\vec{r} \left[\frac{\eta}{2} (\partial_i v_j + \partial_j v_i)^2 \right].$$

(9.15)

Since we regard the fluid and solid similarly as collections of molecules but with different interactions, it is natural to assume that the qualitative form of the energy dissipation at the fluid-solid interface is similar to the viscous dissipation in bulk fluid. That is, if we start with $R_{vis} = \int d\vec{r}[\eta(\partial_z v_x)^2]$ for a shear flow with $\partial_z v_x \neq 0$ only, then for a fluid layer along the solid boundary with thickness Δz normal to the surface, we can write $R_{vis} = \int dS[(\eta/\Delta z)(\Delta v_x)^2]$, where $dS = d\vec{r}/\Delta z$ is the surface differential. Since Δv_x is the relative (tangential) velocity between the fluid layer and the solid boundary, it is precisely what we would call the slip velocity. That directly suggests the form of frictional dissipation rate at the fluid-solid interface to be

$$R_{slip} = \int dS[\beta(v_\tau^{slip})^2],$$

(9.16)

where the slip coefficient β has the dimension of [viscosity]/[length]. Hence a slip length may be defined as $l_s = \eta/\beta$. It should be emphasized that while R_{slip} arises from the assumption of relative slip at the fluid-solid interface; there is no specification of how much slipping there should be. Even an infinitesimal amount of slipping would lead to the form of Eq. (9.16). In both Eqs. (9.15) and (9.16), the rate variable [$\dot{\alpha}_i$ in Eq. (9.14)] is noted to be the fluid velocity.

For dissipations related to inter-diffusion and relaxation, the relevant variable is ϕ. For small perturbations away from the equilibrium (concentration of the two fluid species), there can be dissipation related to the diffusive current in restoring the equilibrium. Since ϕ is locally conserved, it must satisfy the continuity equation

$$\dot{\phi} = \frac{\partial \phi}{\partial t} + \vec{v} \cdot \nabla \phi = -\nabla \cdot \vec{J},$$

(9.17)

where $\vec{J}$ is the diffusive current and also the rate variable. Since the Onsager dissipation function must be quadratic in rates, we thus have

$$R_{diff} = \int d\vec{r} \left[\frac{|\vec{J}|^2}{M} \right],$$

(9.18)

where M is the mobility coefficient. Equation (9.18) is exactly in the form of diffusive dissipation in general. At the fluid-solid interface, ϕ is no longer conserved because diffusive transport normal to the solid boundary is possible ($\partial_n J_n \neq 0$ in general). The only rate variable here is $\dot{\phi}$, hence boundary relaxation would induce a dissipation rate

$$R_{rel} = \int dS \left[\frac{\dot{\phi}^2}{\Gamma}\right],\qquad(9.19)$$

where Γ is a constant related to the boundary mobility coefficient.

For the rate of change of the free energy (second term on the right hand side of Eq. (9.14)), we may write from Eq. (9.2) that since $\delta\phi = \nabla\phi \cdot \delta\vec{x} + (\partial\phi/\partial t)\delta t$, the time variation of the free energy is given by

$$\int d\vec{r}\left[\mu\frac{\partial\phi}{\partial t}\right] + \int dS\left[L\frac{\partial\phi}{\partial t}\right] = \int d\vec{r}\left[\mu(\dot{\phi} - \vec{v}\cdot\nabla\phi)\right] + \int dS\left[L(\dot{\phi} - \vec{v}\cdot\nabla\phi)\right]$$

$$= \int d\vec{r}[\mu\dot{\phi} - \mu\vec{v}\cdot\nabla\phi] + \int dS[L\dot{\phi} - Lv_\tau\partial_\tau\phi]$$

$$= \int d\vec{r}[-\mu\nabla\cdot\vec{J} - \mu\vec{v}\cdot\nabla\phi] + \int dS[L\dot{\phi} - Lv_\tau\partial_\tau\phi]$$

$$= \int d\vec{r}[\nabla\mu\cdot\vec{J} - \mu\vec{v}\cdot\nabla\phi] + \int dS[L\dot{\phi} - Lv_\tau\partial_\tau\phi],\qquad(9.20)$$

where the first term of the last line is obtained from the first term of the previous line through the use of integration by parts, with

$$\int d\vec{r}[\nabla\cdot(\mu\vec{J})] = \int dS[\mu J_n] = 0$$

because $J_n = 0$ at the solid boundary. The last line of Eq. (9.20) may be decomposed into the entropy part and the work part according to general thermodynamics, i.e., $\dot{F} = -T\dot{S} + \dot{W}$, where

$$-T\dot{S} = \int d\vec{r}[\nabla\mu\cdot\vec{J}] + \int dS[L\dot{\phi}],\qquad(9.21)$$

and

$$\dot{W} = \int d\vec{r}[-\vec{v}\cdot(\mu\nabla\phi)] + \int dS[-v_\tau(L\partial_\tau\phi)].\qquad(9.22)$$

Whereas the entropy part reflects the dissipations caused by diffusion and boundary relaxation of the two species of fluid, the work part is seen to be the "elastic" force/stress exerted by the interface on the flow. Here $\mu\nabla\phi$ is the capillary force density and $L\partial_\tau\phi$ the stress of the fluid-fluid interface at the solid boundary. For the stationary state, it is easy to see from the first line of Eq. (9.20) that $\dot{F} = 0$. That is, in steady state the diffusive transport of the fluid-fluid interface must be balanced by the kinematic transport by the flow.

By collecting all the terms, the variational functional A can be written as

$$A[\vec{v}(\vec{r}), \vec{J}(\vec{r}), \dot{\phi}(\vec{r})] = \int d\vec{r} \left[\frac{\eta}{4}(\partial_i v_j + \partial_j v_i)^2\right]$$
$$+ \int dS\left[\frac{\beta}{2}(v_\tau^{slip})^2\right] + \int d\vec{r}\left[\frac{J^2}{2M}\right] + \int dS\left[\frac{\dot{\phi}^2}{2\Gamma}\right]$$
$$+ \int d\vec{r}[\nabla\mu \cdot \vec{J}] + \int dS[L\dot{\phi}] + \int d\vec{r}[-\vec{v} \cdot (\mu\nabla\phi)]$$
$$+ \int dS[-v_\tau(L\partial_\tau\phi)]. \tag{9.23}$$

The above functional should be minimized with respect to the three rates $\left\{\vec{v}, \vec{J}, \dot{\phi}\right\}$, supplemented with the incompressibility condition $\nabla \cdot \vec{v} = 0$. We first minimize with respect to $\vec{J}$. There are only two terms in Eq. (9.23) where the diffusive current appears. The relevant Euler-Lagrange equation is given by

$$\vec{J} = -M\nabla\mu. \tag{9.24}$$

When combined with the continuity Eq. (9.17), Eq. (9.24) directly leads to one of the two equations of motion (9.5). Similar minimization with respect to $\dot{\phi}$ at the solid boundary leads directly to the relaxation boundary condition (9.7c), one of the two boundary conditions denoted as the GNBC.

To minimize with respect to $\vec{v}$, it is necessary to first impose the incompressibility condition by the use of a Lagrange multiplier $\lambda(\vec{r})$, leading to an extra term $\int d\vec{r}[\lambda\partial_i v_i]$. Then the total variation with respect to the fluid velocity can be divided into a group of terms for the bulk:

$$-\eta \int d\vec{r}\left[\partial_j\left(\partial_j v_i + \partial_i v_j\right)\delta v_i\right] - \int d\vec{r}\left[\partial_i\lambda\delta v_i\right] - \int d\vec{r}\left[\mu\partial_i\phi\delta v_i\right] = 0, \tag{9.25}$$

from which we get the Stokes equation

$$-\nabla p + \eta\nabla^2\vec{v} + \mu\nabla\phi = 0, \tag{9.26}$$

where we have identified the Lagrange multiplier $\lambda = -p$. In the presence of inertial forces, the Stokes equation (force $= 0$) is immediately generalized to the Navier-Stokes equation (force $=$ mass $\times$ acceleration) through momentum conservation. That yields Eq. (9.6).

The other group of terms, resulting from the variation of A with respect to (tangential) velocity at the solid-fluid boundary, is given by

$$\eta \int dS[\partial_n v_\tau\delta v_\tau + \partial_\tau v_n\delta v_\tau] + \beta \int dS[v_\tau^{slip}\delta v_\tau] - \int dS[L\partial_\tau\phi\delta v_\tau] = 0, \tag{9.27}$$

from which the boundary condition (9.7d) is obtained. This completes the derivation.

It should not be a surprise to note that Eqs. (9.24), (9.7c) and (9.7d) are examples of linear response, i.e., the response of a dissipative system is linearly proportional to the forcing function. Thus the GNBC is a direct reflection of this general behavior. Also, from our derivation it is clear that Eqs. (9.7c) and (9.7d) constitute a consistent pair. It can be shown that both are important in the transport of the contact line across chemically patterned surfaces.

Also, the GNBC does not exclude the NSBC. In fact, NSBC represents the special case where the slip length is zero, or the slip coefficient β is infinite.

9.6 Implications for Nanoscale Hydrodynamics

Here we compare the numerical predictions of the continuum hydrodynamic system (as given by Eqs. (9.5–9.7)) with MD simulations on molecular-scale fluid(s) flow. In order to carry out meaningful comparisons, the parameters of the system must first be fixed. This can be done easily for the density ρ, viscosity η, slip coefficient β, interfacial tension γ, fluid-fluid interface thickness ξ, and static contact angle θ_s by directly using those measured values from MD simulations. That leaves the two phenomenological parameters M and Γ that theoretically can also be obtained through MD simulations, but here we choose to treat them as parameters to be fixed by fitting one set of MD data, shown in Fig. 9.2. Once fixed, we can predict the velocity profiles, pressure profiles, and interfacial shapes under different flow configurations with no adjustable parameters. Below we show three cases. The first two cases are two-phase immiscible Couette flows between two parallel plates separated by H, with different static contact angles. In the first case, denoted the symmetric case, the fluid-solid interaction on two sides of the contact line (the fluid-solid interfacial tension γ_{fs} in the contuum model) is the same. Hence θ_s is 90°. In the second case, denoted the asymmetric case, the fluid-solid interactions are different on two sides of the contact line, with $\theta_s = 64°$. In the third case, we consider immiscible Poiseuille flow between two parallel plates with separation H and the same static contact angle as the asymmetric case. We will use the length and velocity units from MD simulations. Namely, the velocity unit is $\sqrt{\varepsilon/m}$ and the length unit is σ. Here m denotes the molecular mass, assumed to be the same for both liquids, ε is the energy unit in the Lennard-Jones interaction between the liquid molecules, and σ is the attendant molecular interaction range.

In Fig. 9.3 we show the comparison of the tangential (to the wall) and normal velocity profiles for the symmetric case, at different distances z away from the solid wall. In Fig. 9.4 we show the same comparison for the asymmetric case. In Fig. 9.5 we show the comparison of the interfacial profiles, and in Fig. 9.6 we show both the interfacial profile (9.6a) and velocity (9.6b) comparisons for the Poiseuille flow. In all cases the lines are the numerically evaluated continuum predictions and the symbols are the MD results. Excellent quantitative agreements are seen. In particular, for the tangential velocity profile closest to the solid wall, the near-total slip of the MCL is clearly evident. This is a direct consequence of the uncompensated Young

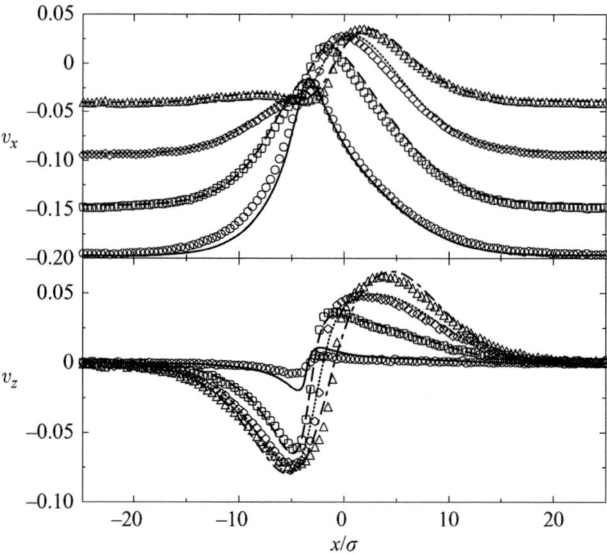

Fig. 9.3 Comparison of the MD (symbols) and continuum (lines) velocity profiles ($v_x(x)$ and $v_z(x)$ at different z levels) for a symmetric case of immiscible Couette flow ($V=0.25(\varepsilon/m)^{1/2}$ and $H = 13.6\sigma$). The profiles are symmetric about the centre plane $z = H/2$, hence only the lower half is shown at $z = 0.425\sigma$ (circles and solid lines), 2.125σ (squares and dashed lines), 3.825σ (diamonds and dotted line), and 5.525σ (triangles and dot-dashed lines)

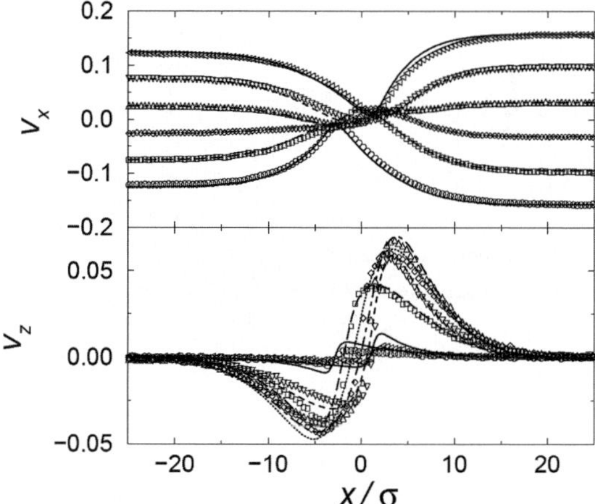

Fig. 9.4 Comparison of the MD (symbols) and continuum (lines) velocity profiles ($v_x(x)$ and $v_z(x)$ at different z levels) for an asymmetric case of immiscible Couette flow ($V=0.2(\varepsilon/m)^{1/2}$ and $H = 13.6\sigma$), shown at $z = 0.425\sigma$ (circles and solid lines), 2.975σ (squares and long-dashed lines), 5.525σ (diamonds and dotted line), 8.075σ (up-triangles and dot-dashed lines), 10.625σ (down-triangles and dashed lines), 13.175σ (left-triangles and solid lines). Although the solid lines are used to denote two different z levels, for each solid line, whether it should be compared to circles or left-triangles is self-evident

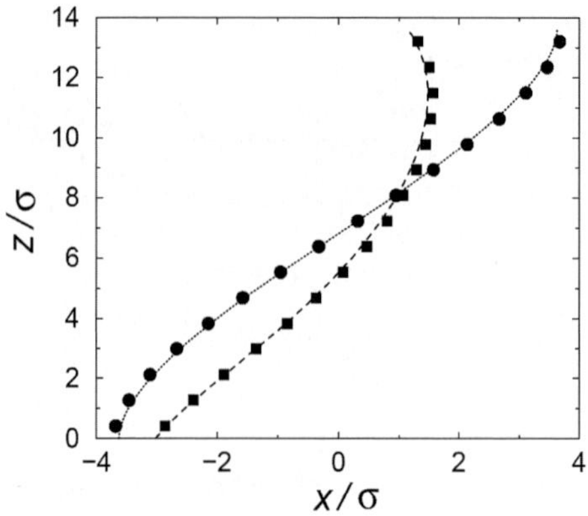

Fig. 9.5 Comparison of the MD (symbols) and continuum (lines) fluid-fluid interface profiles, defined by $\phi = 0$. The circles and dotted line denote the symmetric immiscible Couette flow with $V = 0.25(\varepsilon/m)^{1/2}$ and $H = 13.6\sigma$; the squares and dashed line denote the asymmetric immiscible Couette flow with $V=0.2(\varepsilon/m)^{1/2}$ and $H = 13.6\sigma$

stress (Qian et al., 2003, 2006) (the last term on the right hand side of Eq. (9.7d)), which is a peaked function centered at the contact line. The integral of the uncompensated Young stress (across the contact line) represents the tangential stress that results from the deviation of the dynamic (microscopic) contact angle from the static one. While such deviations are usually small (measured in MD simulations to range from 0.5 to 2°), owing to the relatively large magnitude of γ as compared to the viscous stress (as measured by the capillary number Ca, generally in the range of 0.01–0.1 or even smaller), even a small deviation of the contact angle can imply a

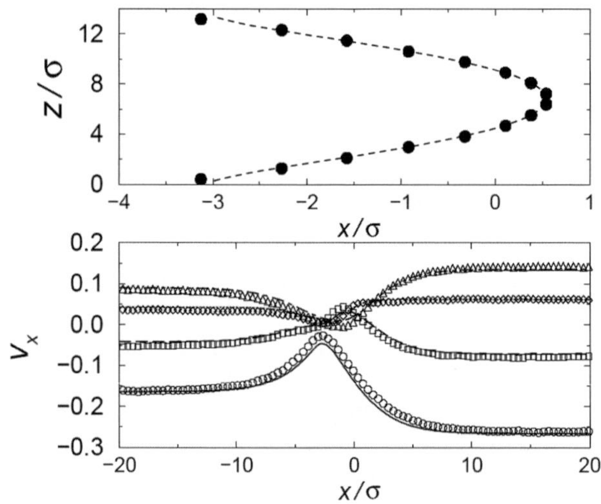

Fig. 9.6 Comparison of the MD (symbols) and continuum (lines) results for an asymmetric case of immiscible Poiseuille flow. An external force $mg_{\text{ext}} = 0.05\varepsilon/\sigma$ is applied on each fluid molecule in the x-direction, and the two walls, separated by $H = 13.6\sigma$, move at a constant speed $V = 0.51(\varepsilon/m)^{1/2}$ in the $-x$-direction to maintain a stationary steady-state interface. (a) Fluid-fluid interface profiles, defined by $\phi = 0$. (b) $v_x(x)$ at different z levels. The profiles are symmetric about the center plane $z = H/2$, hence only the lower half is shown at $z = 0.425\sigma$ (circles and solid line), 2.125σ (squares and dashed line), 3.825σ (diamonds and dotted line), and 5.525σ (triangles and dot-dashed line)

very large tangential projection of the capillary force. This is the basic reason why the Navier boundary condition, which involves just the tangential viscous stress, can not possibly explain the near-total slip of MCL. However, due to the peaked nature of the uncompensated Young stress, away from the contact line the GNBC is identical to the NBC.

An important point to note is that the phenomenon of near-total slip at the MCL is clearly independent of the average fluid velocity. Even at low velocities there is still near-total slip at the MCL. The reason is that while the total tangential stress becomes smaller as the fluid velocity decreases, the absolute magnitude of the slip velocity also decreases. The two are linearly proportional to each other, as stated by Eq. (9.7d). That is, Eq. (9.7d) is in the form of a linear friction law.

Another important piece of information is in regard to the transition behavior from total slip at the MCL to the regions far away, deep in the single phase regime. Is the slip behavior abruptly stopped at some point away from the MCL? Does the amount of relative slip decay exponentially away from the MCL? In fact both of these possibilities were previously considered in the form of ad hoc fixes to the stress singularity (if the NSBC were applied). So what is the true behavior? We

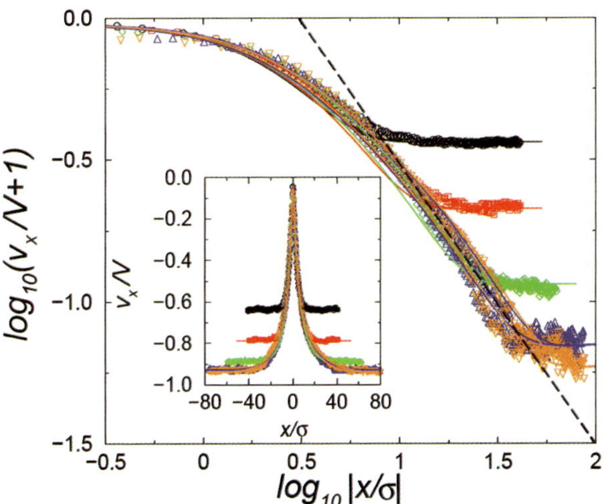

Fig. 9.7 Log-log plot for the slip profiles. Here $v_x/V + 1$ is the scaled slip velocity at the lower fluid-solid interface $z = 0$, and x/σ measures the distance from the MCL in units of σ. The wall is moving at $-V$, hence $v_x/V = 0$ means complete slip and $v_x/V = -1$ means no slip. The v_x profiles were obtained for five symmetric cases of Couette flow, with different values for H but the same value for V ($= 0.05$) and also the same parameters for densities and interactions. The symbols represent the MD results and the solid lines represent the continuum results, obtained for H $= 6.8\sigma$ (black circles and line), H $= 13.6\sigma$ (red squares and line), H $= 27.2\sigma$ (green diamonds and line), H $= 54.4\sigma$ (blue up-triangles and line), H $= 68\sigma$ (orange down-triangles and line). There are two solid curves for each color, one for the slip profile to the left of the MCL and the other to the right of the MCL. The dashed line has the slope of -1, indicating that the $1/x$ behavior is approached for increasingly larger H. For H $= 68\sigma$, the $1/x$ behavior extends from to, where ls was measured to be 2σ. Inset: The scaled tangential velocity v_x/V at $z = 0$, plotted as a function of x/σ

have carried out immiscible Couette flows in increasingly wider channels (Qian et al., 2004). The inset to Fig. 9.7 shows the tangential velocity profiles at the wall. Immediately next to the MCL, there is a small core region where the slip profiles show a sharp decay within a few slip lengths. As the channel width H increases, a much more gentle variation of the slip profiles becomes apparent. In order to reveal the nature of this slow variation, we plot in Fig. 9.7 the same data in the log-log scale. The dashed line has a slope -1, indicating a $1/x$ behavior of the slip profile, where x is the distance from the MCL. Because of the finite H, there is always a plateau in each of the single-phase flow regions, where the constant small amount of slip is given by $2V\ell_s/(H + 2\ell_s)$, which acts as an outer cutoff on the $1/x$ profile. This amount of slip is simply derived from the NBC and the NS equation for uniform shear flow. The continuum results shown in Fig. 9.7 were obtained on a uniform mesh, using the same set of parameter values corresponding to the same local properties in all the five MD simulations. We have extended the MCL simulations, through continuum hydrodynamics, to lower flow rates and much larger systems, using the iterative grid redistribution approach (Ren and Wang, 2000). The $1/x$ partial slip profile is verified to several hundred slip lengths. That means even if the slip length is in the nanoscale, partial slip can extend to mesoscopic distances.

9.7 Concluding Remarks

The most significant implication of our results is the fact that the fluid-solid boundary conditions can be consistently derived from the same statistical mechanic principle that underlies the linear response theories. Making this possible is the viewpoint that both fluid and solid are collections of molecules, so their interfaces, while special, can still be treated within the same theoretical framework. This is the reason why our continuum calculations agree so well with the MD simulations, because the latter inherently treat both fluid and solid as composed of molecules. With the present continuum formulation, it becomes possible now to carry out accurate numerical calculations of nanofluidic/microfluidic phenomena that are beyond the reach of MD simulations.

We note here that the use of Onsager's principle of minimum energy dissipation represents a departure from the traditional theoretical approach of using the kinetic theory to derive the boundary conditions. While the kinetic theory can be useful in the case of rarified gases where the mean free path of collision is large, its application to liquid-solid interface, where the interaction between liquid and solid molecules is strong and the relevant mean free paths very small, can involve difficulties in estimating the effects of the required approximations. Therefore even though the kinetic theory can usually provide a very physical picture of the various processes, it may not be as fruitful as the dissipation point of view in deriving the relevant boundary conditions. It would be most helpful if the two approaches can be combined in some way in the future.

What the resolution of the MCL problem tells us is that the NSBC is at best an approximation. While the partial slip in single phase flows may be quite small, its physical significance can not be minimized. First of all, we can see that without the small partial slip at the fluid-solid interface, there can not be the total slip at the MCL. The two are consistent and can not be separately treated under the framework of minimum energy dissipation principle. Second, the slip coefficient is a parameter whose statistical mechanical underpinning should be similar to that of the bulk viscosity coefficient (Barrat and Bocquet, 1999a,b). Its theoretical and experimental studies remain a fruitful area to be further explored. This is in contrast to the NSBC, which is just a clean statement which lies outside the realm of statistical mechanics.

There are some clear physical implications associated with the GNBC and its derivation framework. Here we would like to mention just two. The existence of a length scale, the slip length, means that the ratio of particle size to the slip length is now an important dimensionless parameter. When that ratio is large (i.e., for large particles), the slip effect is small, and one can approximate the particle transport dynamics (in fluid) by using the NSBC. In particular, the Cauchy-Born principle holds, i.e., the macroscopic deformation is equal to the microscopic deformation. But when the ratio of the particle size to slip length is moderate to small (e.g., for colloids consisting of nanoparticles) then the slip effect can not be neglected, and the Cauchy-Born principle in its traditional form would have to be modified. The fact that some threshold particle size must exist (for the slip effect to be manifest) is not in doubt, since in the limiting case of fluid molecules common sense tells us that relative slip (between the molecules) occurs. A second implication is that when the same framework of minimum energy dissipation principle is applied to complex fluids, there can be order parameters other than the number density. The additional order parameters would on the one hand introduce complexities into the boundary conditions, but on the other hand also imply richness one can explore for targeted purposes, e.g., lubrication. These and other problems are being actively pursued at present.

References

Barrat, J.-L., Bocquet, L., 1999a, Large slip effect at a nonwetting fluid–solid interface. *Physical Review Letters*, **82**, pp. 4671–4674.

Barrat, J.-L., Bocquet, L., 1999b, Influence of wetting properties on hydrodynamic boundary conditions at a fluid/solid interface. *Faraday Discussions*, **112**, pp. 119–128.

Batchelor, G. K., 1967, *An Introduction to Fluid Dynamics*. Cambridge University Press.

Blake, T. D., Haynes, J. M., 1969, Kinetics of liquid/liquid displacement. *Journal of Colloid and Interface Science*, **30**, pp. 421–423.

Briant, A. J., Yeomans, J. M., 2004, Lattice Boltzmann simulations of contact line motion. II. Binary fluids. *Physical Review E*, **69**, p. 031603.

Cahn, J. W., Hilliard, J. E., 1958, Free energy of a nonuniform system. I. Interfacial free energy. *Journal of Chemical Physics*, **28**, pp. 258–267.

Chen, H. Y., Jasnow, D., Vinals, J., 2000, Interface and contact line motion in a two phase fluid under shear flow. *Physical Review Letters*, **85**, pp. 1686–1689.

Cox, R. G., 1986, The dynamics of the spreading of liquids on a solid surface. Part 1. Viscous flow. *Journal of Fluid Mechanics*, **168**, pp. 169–194.

the W parameter is defined as the ratio of the area in the high-momentum region to the total area under the annihilation peak.

$$S, W = \frac{\sum_1 \int_{-\infty}^{\infty} \int_{-\infty}^{\infty} dp_x dp_y \rho(p)}{\sum_{tot} \int_{-\infty}^{\infty} \int_{-\infty}^{\infty} dp_x dp_y \rho(p)}$$

where Σ_1 is over the appropriate energy window for S and W definition, Σ_{tot} is over the full annihilation photopeak.

From the definition, it can be obtained that annihilations with valence electrons are reflected in S parameter, and W parameter contains information from the core electron. The S parameter is more sensitive to the vacancy state of the annihilation site, because the valence electrons have a low momentum that is contributing in the central region of the Doppler-broadening spectra. On the other hand, W parameter is usually used to study the chemical surrounding of the annihilation site as the core electrons have higher momentum.

In this paper, ZnO nanorods were prepared by hydrothermal method on silicon substrates. The nanorods were studied by scanning electron microscopy (SEM), photoluminescence (PL), X-ray photoelectron spectroscopy (XPS), positron annihilation spectroscopy (PAS) and I-V curve measurements before and after annealing in different environments and temperatures. It was found that the yellow emission in hydrothermally grown nanorods can be attributed to the presence of OH groups. The defect emission was strongly affected by the annealing conditions. The origins of the defect emissions were discussed.

10.2 Experimental Details

The ZnO nanorod arrays were prepared from 25 mM aqueous solutions of zinc nitrate hydrate and hexamethylenetetramine. Before the growth of nanorods, silicon substrates were cleaned by sonication in acetone, ethanol, and deionized water, and then dried in an oven. Then, a seed layer was prepared using zinc acetate solution, as described by Greene et al. (2003). After that, the Si substrates with acetate seed layer were placed in the solution at 90 °C for 5 hours. For increasing the aspect ratio of the nanorods, polyethyleneimine was added to the solution (Greene et al., 2005). Zinc nitrate hydrate, hexamethylenetetramine, zinc acetate and polyethyleneimine were obtained from Aldrich. The morphology of the nanorods was examined by scanning electron microscopy (SEM) using a Leo 1530 field emission SEM.

Annealing has been performed in a tube furnace under different conditions. The gas flow rate was 0.1 Lpm, and the pressure was $\sim$ 1 Torr. The optical properties of the annealed ZnO nanorods were investigated using room temperature photoluminescence. The PL spectra were obtained using a fiberoptic spectrometer PDA-512-USB (Control Development Inc.). The excitation wavelength was 325 nm from a He-Cd laser with power of $\sim$ 15 mW.

In order to have more comprehensive study on the defects, PAS and XPS were used to examine the ZnO nanorods before and after annealing. Low-energy positrons were focus to a ~ 1 mm diameter spot on the sample under magnetic guidance in the positron annihilation experiment. The implantation energy of the positrons was varied between 0 and 25 keV. Computer programme VEPFIT was used to fit the experimental data, including positron implantation, positron diffusion, and positron annihilating at different possible sites, for low-momentum S parameter. While XPS spectra were obtained using PAL 102 XPS and the position of the carbon C 1s peak (285.0 eV) was taken as a reference.

Furthermore, ZnO nanorods were grown on the p-type GaN to from a p-n heterojunction. The I-V characteristics of the junction with various annealing conditions were investigated by Keithley 2400 Source Meter.

10.3 Results and Discussions

10.3.1 SEM Characterisation

Figure 10.2 shows the morphology of the as-grown ZnO nanorods. It was observed that the rods were highly oriented. The rod was ~ 500 nm long with diameter range 25–50 nm, while the density of the rods were on the order of several hundred rods per μm^2. In addition, morphology of the rods did not show significant change on different substrates and after annealing.

10.3.2 Photoluminescence

PL spectrum of ZnO nanorods on Si substrates before and after annealing under different conditions are shown in Fig. 10.3. As-grown ZnO nanorods exhibit a UV emission peak and a broad defect emission centre at yellow spectral range. The defect emission was significantly depressed after annealing at $200\,^\circ C$ in all gas flows which were tried in this work (Fig. 10.3a). The results were consistent with the previous report by Hsu et al. (2006). PL spectra for ZnO nanorods prepared by hydrothermal method showed large quenching in the yellow defect emission after

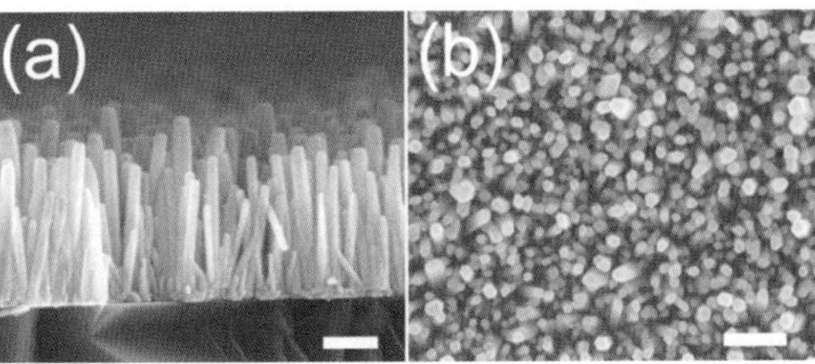

Fig. 10.2 SEM images of ZnO nanorods (a) side view, (b) top view. Scale bar is 200 nm

                                                Djurišić et al.

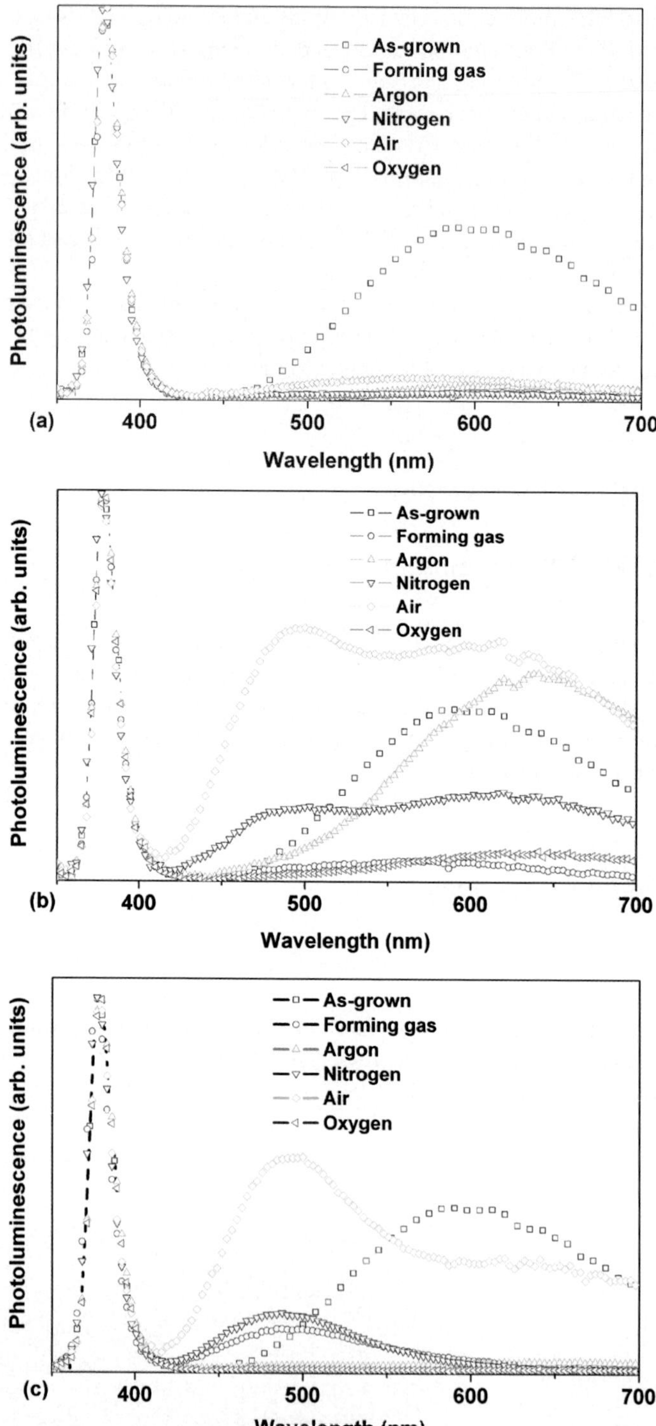

Fig. 10.3 PL spectra of ZnO nanorods annealed in different environments at (a) 200 °C, (b) 400 °C, and (c) 600 °C

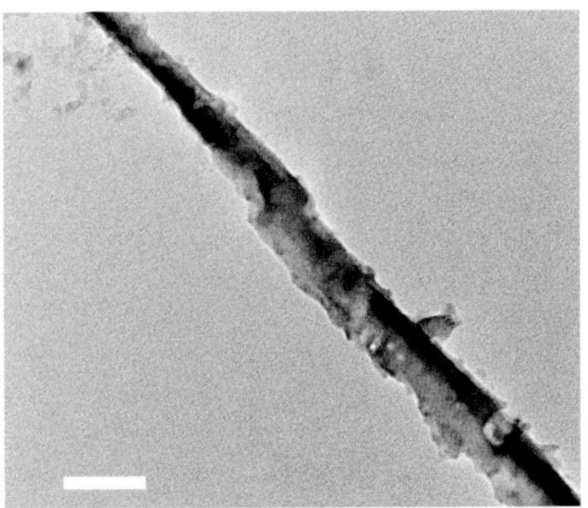

Fig. 10.4 TEM image of ZnO nanorods annealed in forming gas at 600 °C. Scale bar is 50 nm

annealing in reducing environment. This decrease of intensity was likely due to the removal of the hydroxyl groups after annealing (Zhou et al., 2002; Tam et al., 2006). Since the desorption rate of hydroxyl groups peaks at ~ 150 °C, annealing the samples at 200 °C was likely to remove the $Zn(OH)_2$ and water on the surface of the nanorods, corresponding to the depression in the defect emission.

Figure 10.3b and 10.3c show the ZnO nanorods annealed in different environments at 400 °C and 600 °C, respectively. Different visible defect emissions could be observed depending on the annealing environment. Nanorods showed green and orange emissions after annealing in air. Green defect emission could be observed after annealing in N_2 and forming gas (5% hydrogen+95% nitrogen). Surface damage of the ZnO nanorods can be found after annealing in forming gas at higher temperature. The TEM image of the ZnO nanorods annealed at 600 °C is shown in Fig. 10.4. It can be observed that the green emission increases with the surface damage. Relation between the PL and surface damage of ZnO need further study. The origin of the defect emission will be further discussed in Section 10.3.3.

ZnO samples grown by Pal et al. (2006) using low-temperature hydrothermal synthesis also showed similar optical properties. The highest UV to visible ratio of the ZnO samples could be achieved after annealing at ~ 200 °C (the optimum annealing temperature). In order to verify that hydroxyl groups are responsible for the reduction of yellow defect emission, PL was measured on ZnO nanorods at different time after annealed at 200 °C. Figure 10.5a shows the PL spectra of ZnO nanorods stored in air after annealing in Ar at 200 °C. It could be observed that the depressed defect emission reappeared after putting the sample in air. The defect emission increased with an increasing time of storage in air, and became dominant after 96 hours.

ZnO nanorods were coated with 3-aminopropyltrimethoxysilane immediately after annealing at 200 °C for further studies. The inset of Fig. 10.5b shows that similar PL spectrum can be obtained after the nanorods coated with silane. Figure 10.5b shows that almost no change of the PL spectra after storage in air for 96 hours for nanorods coated with silane. The possible reason may be that the hydroxyl groups and water molecules in air were re-adsorbed on the surface of ZnO nanorods and caused the reappearance of the yellow emission. However, ZnO nanorods coated

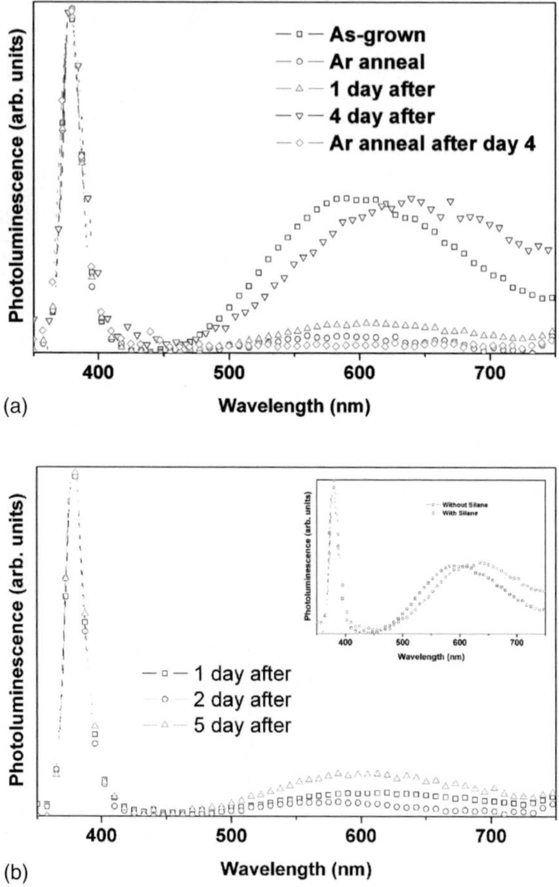

Fig. 10.5 (a) PL spectrum of ZnO nanorods stored in air after annealed at 200 °C. (b) PL spectrum of ZnO nanorods coated with silane stored in air after annealed at 200 °C. Inset: PL spectrum of as-grown ZnO nanorods with and without silane

with silane were inert to the storage environment, thus no significant change in PL spectrum could be found.

10.3.3 PAS and XPS Measurements

In order to obtain more information about the nature of defects in ZnO nanorods, PAS and XPS were used to examine the ZnO nanorods before and after annealing. Table 10.1 shows a summary of the obtained results. In PAS measurement, the effective positron diffusion length ($L_{eff}+$) represents a measure of the probability of positrons being trapped into all kind of defects in a sample, while the S-parameter is only sensitive to certain types of defects, such as Zn vacancy and vacancy cluster or complex in ZnO (Tam et al., 2006). It could be observed that $L_{eff}+$ of the as-grown samples was much shorter than the annealed samples, which was consistent

Table 10.1 Summary on the PAS and XPS results of ZnO nanorods before and after annealing

Annealing conditions	$L_{eff}+$ (nm)	S	O_1/Zn	O_1/O_2	$(O_1+O_2)/Zn$
None	2.8	0.547	0.44	0.69	1.07
200 °C Ar	22	0.559	0.43	0.68	1.07
400 °C Ar	24	0.558	0.44	0.54	1.24
600 °C Ar	28	0.554	0.46	0.55	1.31
200 °C N_2	11	0.560	0.38	0.53	1.11
600 °C N_2	42	0.556	0.42	0.46	1.35

with the low UV to visible ratio in the as-grown samples in PL measurement. The increase of S-parameter upon annealing implied that an increase in Zn vacancy and vacancy cluster. However, the increase of $L_{eff}+$ with increasing annealing temperature showed that the total defect concentration decreased with higher annealing temperature. Therefore, it could only be explained that ZnO annealed in N_2 would have the effect of increasing the concentration of Zn vacancy or vacancy cluster and decreasing the concentration of oxygen vacancy, ionized acceptor or dislocation. The results show that the origin of the green emission exhibits after 600 °C annealing is highly correlated to the Zn vacancy.

Photoelectron spectra of O 1s curves are fitted by two peaks. The lower energy peak is attributed to oxygen ions in the wurtzite ZnO structures, surrounded by Zn atoms, and the higher one is typically assigned to the loosely bound oxygen on the surface of the ZnO. In all of the examined samples, O_1/Zn ratio was lower than 0.5, indicating oxygen deficiency in the bulk of the nanorods. The increase of O_1/Zn and $(O_1+O_2)/Zn$ ratio with higher annealing temperature indicated that there may have an increase of Zn vacancy or a decrease of oxygen vacancy in the samples upon annealing. This result exhibited a good agreement with the PAS measurement. The decrease of O_1/O_2 ratio and increase of $L_{eff}+$ and S parameter suggested that the origin of green emission is involved Zn vacancy or Zn vacancy complexes, as well as surface trapping by chemisorbed oxygen which suggested by van Dijken et al (2000).

10.3.4 I-V Measurement

ZnO nanorods were also grown on the p-GaN on sapphire to perform room-temperature I-V measurement. Figure 10.6 shows the I-V characteristics of heterojunction annealed at various temperatures in Ar flow. It can be observed that all curves show typical diode behaviour. The highest current at the same bias voltage is obtained after annealing in Ar at 200 °C. As-grown sample has the worst performance. It should be noticed that sample with less defect emission in PL seems to have higher current. The correlation between the PL spectrum and the I-V characteristic need to be further studied. However there is evidence that the native defects in ZnO play a very complex role in the performance of the ZnO/GaN diode.

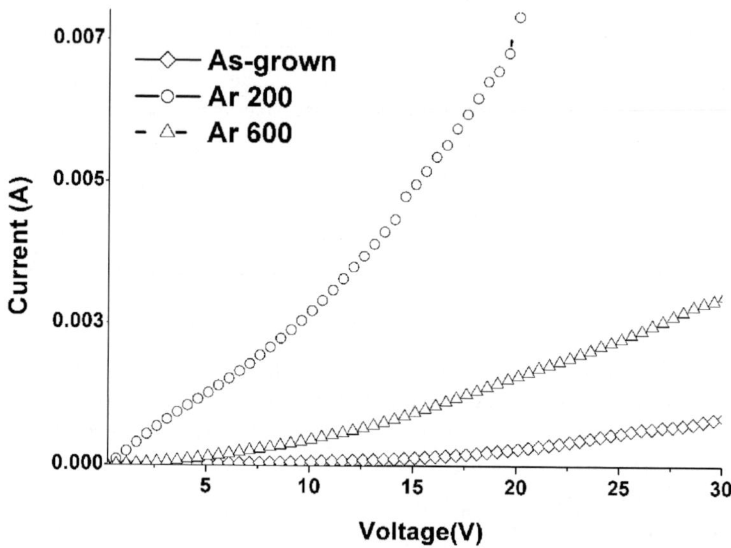

Fig. 10.6 I-V characteristics of as-grown and annealed ZnO nanorods/GaN diodes

10.4 Conclusions

In conclusion, ZnO nanorods have been fabricated by hydrothermal synthesis. As-grown nanorods showed large yellow defect emission. This defect emission could be decreased by annealing the samples at 200 °C in all gases which were tried in this work. However, defect emission reappeared after storing the samples in air. The presence of the OH groups on the hydrothermally grown ZnO nanorods was likely responsible for the yellow emission. The results obtained from PAS and XPS showed that the defects which appeared after annealing at higher temperature possibly involved Zn vacancy. I-V characteristics of the heterojunction of the ZnO/GaN device showed that native defects in ZnO play an important role in the performance.

Acknowledgments The work reported in this paper has been supported by the research grants council of the Hong Kong special administrative region, China (project numbers hku 7008/04p, 7010/05p, and 7019/04p). Financial support from the strategic research theme, university development fund, seed funding grant (administrated by the University of Hong Kong) and outstanding young researcher award are also acknowledged.

References

Borseth, T. M., Svensson, B. G., Kuznetsov, A. Y., Klason, P., Zhao, Q. X. and Willander, M., 2006, Identification of oxygen and zinc vacancy optical signals in ZnO. *Applied Physics Letters*, **89**, p. 262112.

de la Cruz, R. M., Pareja, R., Gonzalez, R., Boatner, L. A. and Chen, Y., 1992, Effect of thermo-chemical reduction on the electrical, optical-absorption, and positron-annihilation characteristics of ZnO crystals. *Physical Review B*, **45**, pp. 6581–6586.

Djurišić, A. B. and Leung, Y. H., 2006, Optical properties of ZnO nanostructures. *Small*, **2**, pp. 944–961.

Djurišić, A. B., Choy, W. C. H., Roy, V. A. L., Leung, Y. L., Kwong, C. Y., Cheah, K. W., Rao, T. K. G., Chan, W. K., Lui, H. F. and Surya, C., 2004, Photoluminescence and electron paramagnetic resonance of ZnO Tetrapod structures. *Advanced Functional Materials*, **14**, pp. 856–864.

Greene, L. E., Law, M., Goldberger, J., Kim, F., Yanfeng, J. C. J., Richard, Z., Saykally, J. and Yang, P., 2003, Low-temperature Wafer-scale production of ZnO nanowire arrays. *Angewandte Chemie International Edition*, **42**, pp. 3031–3034.

Greene, L. E., Law, M., Tan, D. H., Montano, M., Goldberger, J., Somorjai, G. and Yang, P., 2005, General route to vertical ZnO nanowire arrays using textured ZnO seeds. *Nano Letter*, **5**, pp. 1231–1236.

Heo, Y. W., Norton, D. P. and Pearton, S. J., 2005, Origin of green luminescence in ZnO thin film grown by molecular-beam epitaxy. *Journal of Applied Physics*, **98**, pp. 073502(1–6)

Hsu, J. W. P., Tallant, D. R., Simpson, R. L., Missert, N. A. and Copeland, R. G., 2006, Luminescent properties of solution-grown ZnO nanorods. *Applied Physics Letters*, **88**, p. 252103.

Hsu, N. E., Hung, W. K. and Chen, Y. F., 2004, Origin of defect emission identified by polarized luminescence from aligned ZnO nanorods. *Journal of Applied Physics*, **96**, pp. 4671–4673.

Ip, K., Overberg, M. E., Heo, Y. W., Norton, D. P., Pearton, S. J., Kucheyev, S. O., Jagadish, C., Williams, J. S., Wilson, R. G. and Zavada, J. M., 2002, Thermal stability of ion-implanted hydrogen in ZnO. *Applied Physics Letters*, **81**, pp. 3996–3998.

Kaschner, A., Haboeck, U., Martin, S., Matthias, S., Kaczmarczyk, G., Hoffmann, A., Thomsen, C., Zeuner, A., Alves, H. R., Hofmann, D. M. and Meyer, B. K., 2002, Nitrogen-related local vibrational modes in ZnO:N. *Applied Physics Letters*, **80**, pp. 1909–1911.

Lin, Y. J., Tsai, C. L., Lu, Y. M. and Liu, C. J., 2006, Optical and electrical properties of undoped ZnO films. *Journal of Applied Physics*, **99**, p. 093501(1–4).

Ohashi, N., Nakata, T., Sekiguchi, T., Hosono, H., Mizuguchi, M., Tsurumi, T., Tanakaand, J., Haneda, H., 1999, Yellow emission from zinc oxide giving an electron spin resonance signal at g = 1.96. *Japanese Journal of Applied Physics*, **38**, pp. L113–L115.

Ortiz, A., Falcony, C., Hernandez, J., Garcia, M. and Alonso, J. C., 1997, Photoluminescent characteristics of lithium-doped zinc oxide films deposited by spray pyrolysis. *Thin Solid Films*, **293**, pp. 103–107.

Pal, U., Serrano, J. G., Santiago, P., Xiong, G., Ucer, K. B. and Williams, R. T., 2006, Synthesis and optical properties of ZnO nanostructures with different morphologies. *Optical Materials*, **29**, pp. 65–69.

Park, W. I., Kim, D. H., Jung, S. W. and Gyu-Chul Yi, 2002, Metalorganic vapor-phase epitaxial growth of vertically well-aligned ZnO nanorods. *Applied Physics Letters*, **80**, pp. 4232–4234.

Sun, Y., Fuge, G. M. and Ashfold, M. N. R., 2004, Growth of aligned ZnO nanorod arrays by catalyst-free pulsed laser deposition methods, *Chemical Physics Letters*, **396**, pp. 21–26.

Sun, Y., Fuge, G. M. and Ashfold, M. N. R., 2006, Growth mechanisms for ZnO nanorods formed by pulsed laser deposition. *Superlattices and Microstructures*, **39**, pp. 33–40.

Tam, K. H., Cheung, C. K., Leung, Y. H., Djurišić, A. B., Ling, C. C., Beling, C. D., Fung, S., Kwok, W. M., Chan, W. K., Phillips, D. L., Ding, L. and Ge, W. K., 2006, Defects in ZnO nanorods prepared by a hydrothermal method. *Journal of Physical Chemistry B*, **110**, pp. 20865–20871.

Van Dijken, A., Meulenkamp, E. A., Vanmaekelbergh, D. and Meijerink, A., 2000, The Kinetics of the radiative and nonradiative processes in nanocrystalline ZnO particles upon photoexcitation. *J. Phys. Chem. B*, **104**, pp. 1715–1723.

Vanheusden, K., Seager, C. H., Warren, W. L., Tallant, D. R. and Voigt, J. A., 1996, Correlation between photoluminescence and oxygen vacancies in ZnO phosphors. *Applied Physics Letters*, **68**, pp. 403–405.

van Veen, A., Schut, H., De Vries, J., Hakvoort, R. A., Ijpma, M. R., Schultz, P. J., Massoumi, G. R., Simpson, P. J. (eds), 1990, Positron beams for solids and surfaces, AIP Conf. Proc. **218**, AIP, New York, p. 171.

Wagner, R. S. and Ellis, W. C., 1964, Vapor-Liquid-Solid mechanism of single crystal growth. *Applied Physics Letters*, **4**, pp. 89–90.

Wang, Z. L., 2004, Zinc oxide nanostructures: growth, properties and applications. *Journal of Physics: Condensed Matter*, **16**, pp. R829–R858.

Wang, Z. G., Zu, X. T., Zhu, S. and Wang, L. M., 2006, Green luminescence originates from surface defects in ZnO nanoparticles. *Physica E: Low-dimensional Systems and Nanostructures*, **35**, pp. 199–202.

Yi, G.-C., Wang, C. and Park, W. I., 2006, ZnO nanorods: synthesis, characterization and applications. *Semiconductor Science and Technology*, **20**, pp. S22–S34.

Zhou, H., Alves, H., Hofmann, D. M., Kriegseis, W., Meyer, B. K., Kaczmarczyk, G. and Hoffmann, A., 2002, Behind the weak excitonic emission of ZnO quantum dots: $ZnO/Zn(OH)_2$ core-shell structure. *Applied Physics Letters*, **80**, pp. 210–212.

Part III
Nano Materials Design and Synthesis

Chapter 11
Towards Surface Science Studies of Surfaces Formed by Molecular Assemblies Using Scanning Tunneling Microscopy

Chen Wang* and Shengbin Lei

11.1 Introduction

The adsorption of molecules on surfaces are the central issues for classical surface science studies since the beginning of this important field. It could be noted that the fundamental investigations have been dedicated to the surfaces of metals, semiconductors and oxides which have great implications in catalysis, microelectronic devices etc. With the fast evolving research field of constructing molecular architectures with expectations on various functions, the knowledge at single molecule level on the interacting mechanisms are keenly needed. The development of the molecular nanostructures has also ushered in an interesting studying topic relating to the interfaces between dissimilar molecular species. Studies on heterogeneous molecular interfaces could have multiple implications from understanding of fundamental aspects of inter-molecular interactions, to the working mechanisms of molecular devices. This category of studies could also form a natural extension of the classical surface science principles for metals, semiconductors and oxides. To this aspect, scanning tunnelling microscopy (STM), with the demonstrated resolution capability at single molecule level, could provide unique approaches to reveal interactions between dissimilar molecules.

Numerous STM studies have helped gain important insight on the interactions among dissimilar molecules adsorbed on supporting surfaces, such as those in the molecular assemblies prepared under various conditions. In a less noted aspect, the STM can also provide direct evidence on the molecular adsorptions occurred on top of the close packed molecular layers serving as the support surfaces in the way exactly as traditional pristine surfaces of metals, semiconductors and oxides. The experimental foundations enabling such studies are due to the known STM resolution capability within one nanometer or so in the surface normal direction on conductive surfaces. This detection range is well suited for construction of two or even more molecular layers of similar or dissimilar species on top of the support conducting solid surfaces. From molecular adsorption's perspective, the similarities of the investigations on such molecular surfaces with traditional solid surfaces are

---

* National Center for Nanoscience and Technology, Beijing, 100080, China

Z. Tang, P. Sheng (eds), *Nano Science and Technology.*
© Springer 2008

conceptually identical, such as the nature of adsorption (either chemical or physical adsorptions), adsorption induced reconstruction, adsorption selectivity, diffusion behavior etc. In essence, such molecular surfaces represent the category of organically functionalized surfaces with the properties dominated by the constituting molecular building units. The main differences of the molecular surfaces from the traditional solid surfaces are originated with the chemical and structural heterogeneity of the molecular building units, which also offers a rich ground for systematic investigations. The results on the surface properties of the molecular surfaces should surely help enrich the scope of surface science studies by incorporating the tremendous synthetic capabilities of organic-molecule-based building blocks, and also provide much needed knowledge critical for achieving controlled designs of molecular functional devices. This work is intended to present some results showing the possible effects of molecular adsorption on above specified molecular surfaces using STM. These results are derived from a limited number of systems and could provide useful references for broadened scope of studies.

11.2 Construction of Molecular Surfaces for Heterogeneous Molecular Adsorptions

The realization of molecularly decorated surfaces can be directly achieved by two-dimensional assembling of molecular building units. The general principles for assembling the organic molecules are themselves being extensively explored and provide important guidelines for designing molecular surfaces with periodically distributed functional groups. A wide range of molecular assemblies have been analyzed and the revealed results stimulated much interest in the driving mechanisms behind molecular architectures (Rabe and Buchholz, 1991; Claypool et al., 1997; Cyr et al., 1996; Gimzewski and Joachim, 1999; De Feyter and De Schryver, 2003). The scope of undertaken investigations includes various conditions (ultra-high vacuum, ambient, solutions) and surfaces (metal, graphite, silicon, etc.).

The formation of well-ordered two-dimensional assembling structures is jointly affected by intermolecular interactions and molecule-substrate interactions (Rabe and Buchholz, 1991; Claypool et al., 1997; Cyr et al., 1996; Hentschke et al., 1992). When multi-component mixtures are involved in the self-assembly process, the weak interactions between self-assembled molecules are very important in the assembling of supramolecular architectures. It should be noted that hydrogen bond plays an important role in building supramolecular nanostructures with its advantage of selectivity and directionality, (Lei et al., 2001; De Feyter et al., 2005) and is widely used and studied in molecular self-assemblies. Examples for the application of hydrogen bond could be found in the construction of the 2D networks of carboxyl derived porphyrin and copper phthalocyanine (CuPc), (Lei et al., 2001) and also the coded assembling of donor-acceptor supramolecular assembly of perylene bisimide and diamino triazine derivative. (De Feyter et al., 2005) Other important interacting forces, i.e. van der Waals and electrostatic interactions, are also prevalent in constructing molecular surfaces or networks. As an example, it is well-known that ordered two-dimensional monolayers of alkanes and alkane derivatives can be formed through physisorption on highly oriented pyrolytic graphite (HOPG) (Rabe and Buchholz, 1991; Hentschke et al., 1992). The crystallization effect of linear alkane chains due to van der Waals interactions is dominant in such systems. The functional terminal groups of the alkane derivatives are periodically arranged within the alkane lamella templates and form a chemically decorated molecular surface.

11.3 Selective Adsorption Behavior of Molecular Surfaces

The above-mentioned molecular surfaces have been shown to have a number of effects, such as selective adsorption, modulated assembling, immobilization effect of adsorbates, etc. With rich variety of functional groups and intermolecular interactions, investigations on the novel effects of assembled molecular architectures should be highly rewarding.

The reported work have identified that both physical and chemical adsorptions could occur on molecular surfaces, depending on the functional groups of the constituting molecules of the surface and adsorbates. A readily available example of molecular surface can be found with linear alkane derivatives (such as stearic acid, 1-octadecanol and 1-iodooctaecane) which are known to form parallel lamella patterns on graphite surface in ambient conditions. When pre-covered on the support surfaces (such as graphite or metals), the molecular lamella structures introduce inherently the heterogeneous adsorption sites and anisotropic diffusion barriers in association with the functional groups. The presence of heterogeneous adsorption sites could result in selective adsorption of single molecules. The adsorbed species would also experience the anisotropic diffusion barrier and organize in a restricted manner.

It was observed that single or clusters of copper phthalocyanine (CuPc) molecules could be stably adsorbed on alkane derivatives serving as the support surface (Xu et al., 2000; Lei et al., 2004a). In the case of single CuPc molecule adsorption on the lamella structure formed by stearic acid and 1-iodooctadecane, the Pc molecules were found to be exclusively located on top of the alkane part of stearic acid and 1-iodooctadecane lamellae, and the location of functional groups are not the preferred adsorption sites. Possibly due to the coherence in geometry, pairs of CuPc molecules in close contact side-by-side is preferentially observed to lie along the alkane chains (Fig. 11.1a).

In another type of molecular surface consisted of tridodecyl amine (TDA) lamellae, both physical and chemical adsorptions were identified. Physical adsorption behavior was observed for CuPc molecules that are again exclusively adsorbed at the alkane part of the tridodecyl amine (TDA) lamellae. At low adsorbate concentration molecule pairs of CuPc appear again on the TDA template (Fig. 11.1b). However, different from the case on 1-iodooctadecane, the CuPc molecules in the molecular pair are separated by the amine groups. The distance between CuPc centers is measured to be nearly 1 nm. Though the dominating interaction between adsorbates and molecular surface is van der Waals and therefore can be characterized by physical adsorption, the formation of molecular pairs separated by the amine groups indicate the interaction between the CuPc molecule and the amine group plays some role in

Fig. 11.1 Isolated single CuPc molecule and molecular pairs observed on top of the alkane part of 1-iodooctadecane (a) and TDA (b). A double-molecule band of CuPc modulated by the TDA layer has been observed as the increase of CuPc coverage (c). Image condition, a: $-772\,\mathrm{mV}$, 761 pA; b: 990 mV, 320 pA; c: 724 mV, 876 pA

the formation of such CuPc dimers. It was further revealed that with the increase of adsorbate coverage, a complete overlayer of CuPcs will be modulated by the periodicity of the underneath TDA lamellae (Lei et al., 2004b). Such modulation effect demonstrates the impact of periodicity of molecular surface (in this specific study, alkane derivative lamellae) on the adsorption, diffusion and assembling of adsorbed molecules. One thing worth noticing is the submolecular resolution obtained on such isolated single molecules and clusters. Normally for STM imaging on the solid/gas or solid/liquid interface under ambient conditions submolecular resolution could only be obtained on well ordered crystalline domains. The high resolution achieved on the isolated molecules on the modulated surface indicates high stability of these molecules. This stability could probably be attributed to the increased diffusion barrier for the CuPc molecules on the modulated surface.

The example of chemical adsorption can be found in the hydrogen bond assisted adsorption of benzoic acid molecules exclusively at the location of amine groups in the TDA assembly (Lei et al., 2003). TDA forms well ordered lamellae structure when deposited on the surface of graphite (Fig. 11.2a). High-resolution STM image reveals that TDA molecules adsorb with their alkyl chains parallel to the basal plane

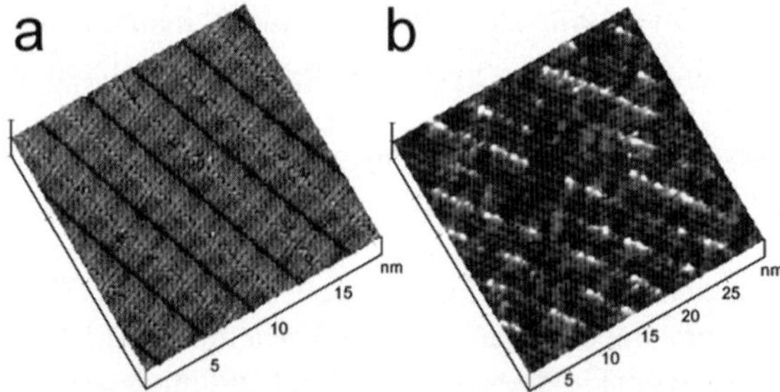

Fig. 11.2 TDA assembly observed without (a) and with (b) the presence of benzoic acid. The amine groups appear with dark contrast in the image without benzoic acid, but when benzoic acid is added into the solution, isolated bright spots appear in the position of the amine groups. Image condition: a: 1.0 V, 426 pA; b: 1.04 V, 14.6 pA

of graphite with a 90° angle to the lamella axis. The amine groups show reduced contrast with respect to the alkane part of the molecule. When benzoic acid was added, bright spots with diameter of about 0.6 nm appears along the center of the TDA lamellae (Fig. 11.2b). The diameter of these bright spots is in agreement with the dimension of single benzene rings. Compared with the observed lamellae in the uncovered area, these bright spots could be attributed to benzoic acid molecules over the sites of the amino groups in the TDA lamellae, possibly through the $O-H-N$ hydrogen bond. It was also observed that the adsorbed benzoic acids on the surface could be easily removed by tip scanning, indicating the linkage between the benzoic acid and amine group is weak. Another example of chemical adsorption can be found in the urea molecules adsorption on the lamella of double-alkyl amino acid (Hoeppener et al., 2003), in which the unprotected amino acid groups were found to be the preferential adsorption sites for urea molecules.

11.4 Summarizing Remarks and Future Perspectives

The above discussed examples illustrate the possibilities of adopting molecular surfaces as an important category of support medium. They are ideally suited as model systems for mechanistic studies of molecular based devices. The structural resolution capability of STM can be optimized, in combination with molecular self-assembling processes, to yield insightful knowledge on the interface between dissimilar molecular layers. It is conceivable that with the tunneling spectroscopic capability of STM, one could further pursue the electronic characteristics of the heterogeneous molecular interfaces. In addition, the vast possibility of organizing chemical groups in 2D molecular assemblies would also provide much unexplored novel effects of heterogeneous molecular interfaces.

References

Claypool, C. L., Faglioni, F., Goddard III, W. A., Gray, H. B., Lewis, N. S., Marcus, R. A., 1997, *J. Phys. Chem. B* 101, 5978.

Cyr, D. M., Venkataraman, B., Flynn, G. W., Black, A., Whitesides, G. M., 1996, *J. Phys. Chem.* 100, 13747.

De Feyter, S., De Schryver, F. C., 2003, *Chem. Soc. Rev.* 32, 139.

De Feyter, S., Miura, A., Yao, S., Chen, Z., Wurthner, F., Jonkheijm, P., Schenning, A. P. H. J., Meijer, E. W., De Schryver, F. C., 2005, *Nano letts.* 5, 77.

Gimzewski, J. K., Joachim, C., 1999, *Science* 283, 1683.

Hentschke, R., Schurmann, B. L., Rabe, J. P., 1992, *J. Chem. Phys.* 96, 6213.

Hoeppener, S., Wonnemann, J., Chi, L. F., Erker, G., Fuchs, H., 2003, *Chem Phys Chem.* 4, 490.

Lei, S. B., Wang, C., Wan, L. J., Bai, C. L., 2003, *Langmuir* 19, 9759.

Lei, S. B., Wang, C., Yin, S. X., Wang, H. N., Xi, F., Liu, H. W., Xu, B., Wan, L. J., Bai, C. L., 2001, *J. Chem. Phys. B.*, 105, 10838.

Lei, S. B., Yin, S. X., Wang, C., Wan, L. J., Bai, C. L., 2004a, *J. Phys. Chem. B* 108, 224.

Lei, S. B., Wang, C., Wan, L. J., Bai, C. L., 2004b, *J. Phys. Chem. B* 108, 1173.

Rabe, J. P., Buchholz, S. 1991, *Science* 253, 424.

Xu, B., Yin, S. X., Wang, C., Qui, X. H., Zeng, Q. D., Bai, C. L., 2000, *J. Phys. Chem. B* 104, 10502.

Chapter 12
Electronic Transport Through Metal Nanowire Contacts

Y. H. Lin, K. J. Lin, F. R. Chen, J. J. Kai and J. J. Lin[*]

12.1 Introduction

Nanoscale structures have recently become the subject of intense theoretical and experimental investigations because of their importance for both fundamental researches and potential industrial applications. Among the various nanostructures, self-assembled quasi-one-dimensional metallic nanowires are of particular interest, primarily due to their rich and fascinating electrical properties (see, for example, Lin et al., 2003; Liu et al., 2007). To explore how electrons transport through these quasi-one-dimensional systems, nanofabrication techniques such as the electron-beam lithography and the focused ion beam deposition are often employed to this end. Nevertheless, it is known that fabrication of reliable electronic contacts to individual nanostructures has been a nontrivial issue, since very often an imperfect contact inevitably forms and may possess a non-negligible temperature dependent contact resistance, R_c, which in turn can seriously complicate the measurements. In the cases involving semiconductors, the contact between a metal electrode and a semiconductor nanowire usually forms a Schottky barrier, whose properties have been extensively studied (Ip et al., 2004). A different situation involving heavily doped semiconductor nanowires has also been discussed (Gu et al., 2001). On the other hand, the electrical properties of the contact between a metal electrode and a metal nanowire have not yet been much addressed in the literature. In the case of metal nanowires, since the as-grown nanowires may readily be covered by a layer of some insulators (oxidations, contaminations, amorphous coating, etc.) of a few nm thick, the underlying physics of how electron waves transmit through such a nanoscale interface is of fundamental interest and urgent industrial concerns.

In this work, we report the temperature behaviour down to liquid-helium temperatures of high-resistance electronic contacts formed at the interfaces between lithographic-patterned submicron electrodes and single metal nanowires. The contact resistances $R_c(T)$'s were determined through the electrical measurements on individual metal nanowires by using the 4-probe, together with the 2-probe, configurations as schematically depicted in Fig. 12.1a–c. In the 4-probe configuration

---

[*] Institute of Physics and Department of Electrophysics, National Chiao Tung University, Hsinchu 30010, Taiwan e-mail: jjlin@mail.nctu.edu.tw

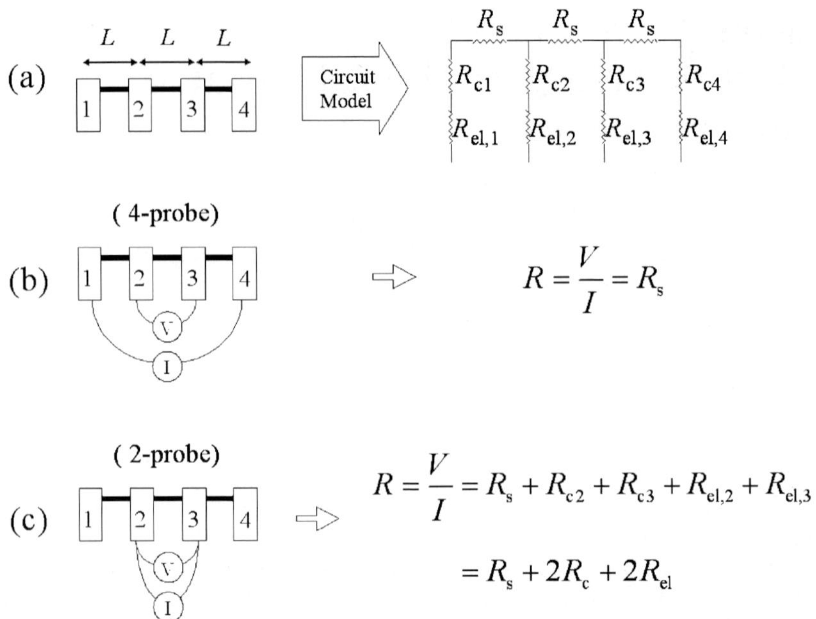

Fig. 12.1 (a) Schematic diagram for a single nanowire (thick bar) with four electronic contacts on it and the equivalent circuit model. R_s denotes the resistance of each segment of the nanowire (i.e., the sample resistance). R_{c1}, R_{c2}, R_{c3} and R_{c4} denote the electronic contact resistances between the nanowire and the submicron electrodes 1, 2, 3 and 4, respectively. $R_{el,i}$ denotes the submicron-electrode resistance of the ith electrode. Four-probe (b) and two-probe (c) measurement configurations, and the corresponding measured resistances. In (c), the approximations $R_{c2} \approx R_{c3} \equiv R_c$ and $R_{el,2} \approx R_{el,3} \equiv R_{el}$ are made (see text)

(Fig. 12.1b), the *intrinsic* nanowire sample resistance, R_s, can be accurately determined. In the 2-probe configuration (Fig. 12.1c), the measured resistance is the sum of R_s plus the two contact resistances and the two submicron-electrode resistances, i.e., $R(T) = R_s(T) + 2R_c(T) + 2R_{el}(T)$, where R_{el} is the submicron-electrode resistance. The prefactor 2 is introduced to denote that there are two similar contacts (submicron electrodes) in series in the 2-probe method. Here we have assumed the simple case that the two electronic contacts are (almost) symmetric, because they were fabricated simultaneously under the same conditions on the same nanowire. (Experimentally, the resistance of each electronic contact can be quantitatively extracted by further applying the 3-probe measurement configuration in addition to the 2-probe measurement configuration, as will be described elsewhere (Lin et al., 2007).) In the case of high-resistance contacts, i.e., $R >> R_s$ and $R >> R_{el}$, the R_c may dominate the measured resistance and the approximation $R \approx 2R_c$ is valid for the whole range of experimental temperature. The measured magnitude and temperature behaviour of R thus faithfully reflect the magnitude and temperature behaviour of R_c. As a result, $R_c(T)$ can be quantitatively inferred. In the present work, single-crystalline RuO_2 nanowires, which exhibit metallic conductivity comparable with that of normal metals (Ryden et al., 1970; Lin et al., 2004) and at the same time possess excellent thermal and chemical stability (Trasatti and Lodi, 1981; Trasatti, 1991), were chosen as our sample system to substantiate this idea.

12.2 Experimental Method

Self-assembled single-crystalline RuO_2 nanowires were synthesized by the thermal evaporation method based on the vapour-liquid-solid growth mechanism (Wagner and Ellis, 1964). The as-grown nanowires were characterized using field-emission scanning electron microscopy (FE-SEM), transmission electron microscopy (TEM), x-ray diffraction (XRD), and x-ray energy dispersive spectroscopy (XR-EDS), as described previously (Liu et al., 2007).

Electrical contacts onto the nanowires were fabricated as described below. A Si substrate with a $\approx$ 200-nm thick SiO_2 layer grown on top was first photo-lithographically patterned with Cr/Au ($\approx$ 10/60 nm) "macro-electrodes" using bi-layer photoresist process to create reverse-slope resist sidewall profiles. Several droplets of dispersed alcoholic solution containing RuO_2 nanowires were dropped on the substrate. Following the standard electron-beam lithography process, a single nanowire was positioned with SEM, and "submicron electrodes" contacting the nanowire were generated by thermal evaporation of Cr/Au ($\approx$ 10/90 nm). Fine Cu leads were attached to the macro-electrodes with silver paste, and the substrate was thermally anchored to the sample holder on a standard ^{4}He cryostat. Depending on the different probe configurations employed, the resistances were measured by two different manners. For the 4-probe method, a Linear Research LR-700 AC resistance bridge was applied. For the 2-probe method, the resistances were measured by utilizing a Keithley K-6430 source meter as a current source and a K-182 nano-voltmeter, in which the DC current-reversal method was adopted so that any existing thermoelectric voltages along the measurement loop were cancelled. We notice that for all of the $R(T)$ curves reported in this work, a measurement current of 10 nA was applied and the current-voltage linearity had been checked and ensured at various temperatures.

12.3 Results and Discussion

To ensure the metallic nature of our single-crystalline RuO_2 nanowires, the resistivities ρ as a function of temperature from 300 K down to liquid-helium temperatures for individual as-grown nanowires were first determined with the 4-probe measurement configuration. Several RuO_2 nanowires with different diameters ranging from a few tens to about 150 nm have been studied. Figure 12.2 shows the result for a representative nanowire with a diameter of $\approx 55 \pm 5$ nm. (The diameter of the nanowire was determined by AFM.) Clearly, the nanowire reveals electrical-transport characteristic of a typical metal, i.e., the resistivity decreases with decreasing temperature. Inspection of Fig. 12.2 indicates that the resistivity $\rho(300\,\mathrm{K}) \approx 240\,\mu\Omega$ cm

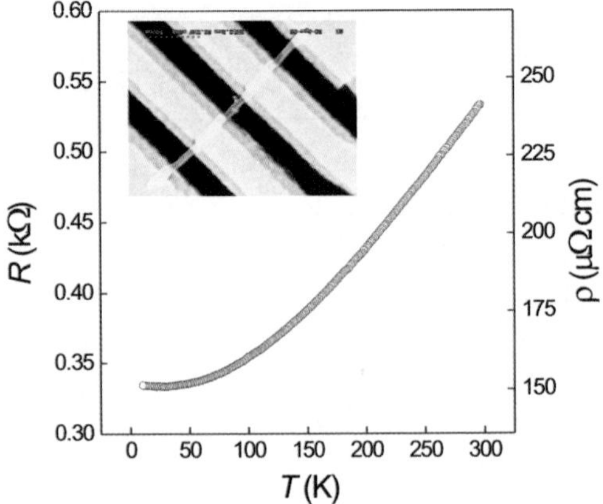

Fig. 12.2 Resistivity as a function of temperature for a RuO_2 nanowire with diameter of $\approx 55 \pm 5$ nm. The inset shows the SEM image, where the scale bar is $1\,\mu$m. Notice that the length of the nanowire used for the evaluation of the resistivity from the measured resistance is taken to be the distance between the middles of the two voltage probes

in this nanowire is relatively higher than that ($\approx 85 \pm 35\,\mu\Omega$ cm) in bulk single crystals (Ryden et al., 1970; Lin et al., 2004). On the other hand, the resistivity ratio $\rho(300\,\text{K})/\rho_0 \approx 1.6$ for this nanowire is considerably lower, as compared with the corresponding bulk value (≈ 100–1000, depending on the quality of the crystal (Ryden et al., 1970; Lin et al., 2004)), where ρ_0 is the residual resistivity. In fact, we have found $\rho(300\,\text{K}) \approx 150$–$280\,\mu\Omega$ cm and $\rho(300\,\text{K})/\rho_0 \approx 1.5$–$1.7$ in a number of RuO_2 nanowires. Such high resistivity values and low resistivity ratio values imply the presence of a high level of defects (most likely, point defects) in the as-grown nanowires. This observation is in sharp distinction to the conclusion drawn from conventional materials characterization techniques, such as XRD and high-resolution TEM (Liu et al., 2007), which often suggested very high quality atomic structure. Similar phenomenon had also been found in single-crystalline metallic NiSi nanowires (Wu et al., 2004). This observation suggests that the electrical-transport measurements over a wide range of temperature can serve as a powerful probe for the presence of microscopic defects in nanostructures. For instance, the electron mean free path, which is highly sensitive to the existence of any level of randomness in the crystal, can be quantitative determined from the $\rho(T)$ measurements. On the other hand, with the metallic nature of our individual RuO_2 nanowires being established, we are safe to apply these nanowires to quantitatively explore the electronic contact resistances $R_c(T)$ in lithographic-contacting nanodevices, employing the 2-probe measurement configuration discussed above.

Ideally, the lead resistances and the electronic contact resistances in electrical measurements must be small to minimize thermal noises. In our case, the electronic contact resistances formed between the electron-beam lithography patterned electrodes and the RuO_2 nanowires normally fall between several tens and several hundreds Ω, and are not much dependent on temperature. However, highly resistive electronic contacts with room temperature resistances of order several kΩ or higher may also be obtained in many fabrications. Figure 12.3 shows typical temperature behaviour of R_c for such a high-resistance contact from 300 K down to liquid-helium temperatures, as determined from the 2-probe method. In strong contrast to the 4-probe result shown in Fig. 12.2, now the measured resistance reveals semiconducting or insulating behaviour, i.e., the resistance increases rapidly with decreasing temperature. As discussed, in this case $R(T) \approx 2R_c(T)$, since the two submicron Cr/Au electrodes and the RuO_2 nanowire have resistances much lower than the measured value. (The total Cr/Au electrode resistance is typically $\approx$ a few tens Ω, and the RuO_2 nanowire resistance is typically $\approx$ a few hundreds Ω.) The large contact resistance ($\sim$ several tens kΩ) may result from a thin, dirty insulating layer incidentally formed at the interface between the submicron electrodes and the nanowire. This insulating layer could be the lightly contaminated or oxidized metals introduced during the electrode evaporation, the amorphous coating resulting from the complex growth process, the vacancies caused by dramatic surface roughness near the contact region, or the breaking induced by tensile stress.

In view of the fundamental researches and industrial applications of nanoscale structures, it is very important to understand the behaviour and the underlying physics of the electronic contacts formed between interconnects and the metal nanostructure. The inset to Fig. 12.3 shows a plot of a representative variation of $\log R$ with T^{-1} for one of our contact resistances. This figure indicates that the simple thermal activation conduction (the straight solid line) is only responsible near room temperatures. As the temperature reduces from room temperature, the resistance does not increase as fast as would be expected from the thermally activated process. At liquid-helium temperatures, the resistance appears roughly

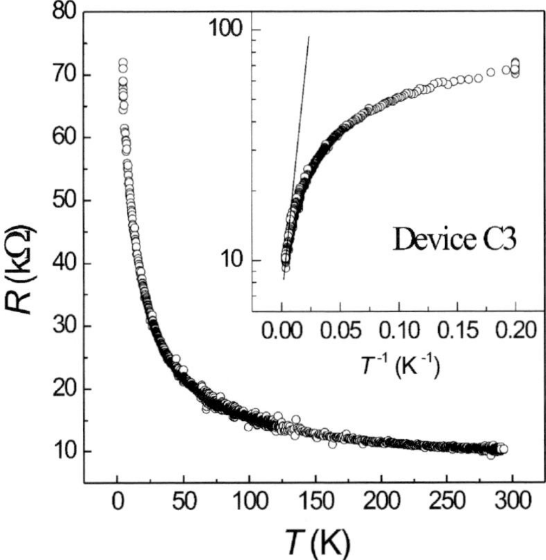

Fig. 12.3 Resistance as a function of temperature for two high-resistance contacts in series, as determined from the 2-probe measurement configuration on the nanowire device C3 (see Table 12.1). The inset shows a plot of $\log R$ versus T^{-1} for the same nanowire device. The straight solid line indicates the thermal activation conduction

constant, i.e., independent of temperature. Such a temperature independent resistance at liquid-helium temperatures can signify a conduction mechanism characteristic of simple elastic tunnelling. Indeed, quantitative analysis (see below) indicates that the overall temperature behaviour of the contact resistance can be well interpreted in terms of a tunnelling conduction model. Previously, in order to explain the electrical-transport properties in certain classes of granular metal-dielectric composites, Sheng and co-workers (1978) and Sheng (1980) have proposed a thermally "fluctuation-induced tunnelling" model, where the thermal effects arising from the capacitance C of a *small* junction formed between two *large* metal grains was considered.

According to Sheng and co-workers (1978) and Sheng (1980) the temperature dependent resistance for small applied electric fields across a single small junction can be expressed as

$$R(T) = R_0 \exp\left(\frac{T_1}{T_0 + T}\right) \tag{12.1}$$

where R_0 is parameter which depends only weakly on temperature, and T_1 and T_0 are characteristic temperatures defined as

$$T_1 = \frac{8\varepsilon_0}{e^2 k_B}\left(\frac{A V_0^2}{w}\right) \tag{12.2}$$

and

$$T_0 = \frac{16\varepsilon_0 \hbar}{\pi (2m)^{1/2} e^2 k_B}\left(\frac{A V_0^{3/2}}{w^2}\right) \tag{12.3}$$

where ε_0 is the vacuum permittivity, $\hbar$ is the Planck's constant divided by 2π, and m is the electron mass. In Eq. (12.1), T_1 can be regarded as a measure of the energy required for an electron to cross the potential barrier between the two conducting regions, and T_0 is the temperature below which the fluctuation effects become insignificant, since, at $T \ll T_0$, Eq. (12.1) is temperature independent and reduces to an expression for the expected simple elastic tunnelling. In the derivation of Eq. (12.1), the conduction was first modelled (Sheng et al., 1978) as the tunnelling of electrons through a single potential barrier of width w, height V_0, and junction area A. (Here A is the size at the point of the two large conducting regions' closest approach.) If A is small enough, it was found (Sheng et al., 1978; Sheng, 1980) that the potential barrier seen by the electrons could be effectively narrowed and lowered by the thermal voltage fluctuations ($\approx (k_B T/C)^{1/2}$, where k_B is the Boltzmann constant) across the insulating gap due to the small effective capacitance of the junction. Such a potential-barrier modulation effect greatly influences the tunnelling probability in the low temperature limit, and consequently introduces a characteristic temperature behaviour to the normally temperature independent tunnelling conductivity (i.e., the elastic tunnelling regime).

In the case of granular composites of macroscopic sizes, it was then argued (Sheng, 1980), via the effective-medium theory, that in a network of independently fluctuating tunnel junctions with different values of T_1 and T_0, the conductivity of the network could still be well described in terms of a single junction with a representative set of T_1 and T_0.

Figure 12.4 shows a plot of our experimental results in double logarithmic scale for four representative NW devices having high contact resistances (as determined from the 2-probe method). The symbols are the experimental data and the solid curves are the least-squares fits to Eq. (12.1), with R_0, T_1 and T_0 as the adjusting parameters. Inspection of Fig. 12.4 clearly indicates that the Eq. (12.1) can well describe the overall temperature behaviour for a wide range of temperature between 2 and 300 K. The values of T_1 and T_0 can then be reliably extracted. Furthermore, by using SEM and/or AFM, we can directly measure the diameter and the length of our nanowires, as well as the width of the relevant submicron Cr/Au electrodes overlying the nanowire. Therefore, the junction area A which appeared in Eqs. (12.2) and (12.3) is *independently* determined. (In this work, the junction area A is given by the product of the nanowire diameter and the width of the overlying submicron electrode.) With the values of T_1, T_0 and A being determined, the microscopic parameters characterizing the electronic contacts, i.e., the width w and height V_0 of the potential barrier, may then be inferred. Our experimental values of the relevant parameters are listed in Table 12.1. Notice that, in Table 12.1, the measured resistance $R(300\,\mathrm{K})$ for each nanowire device is at least an order of magnitude higher than the intrinsic resistance of the nanowire, $R_\mathrm{s}(300\,\mathrm{K})$, justifying our approximation $R(T) \approx 2R_\mathrm{c}(T)$. Furthermore, the resistance ratio $R(T)/R_\mathrm{s}(T)$

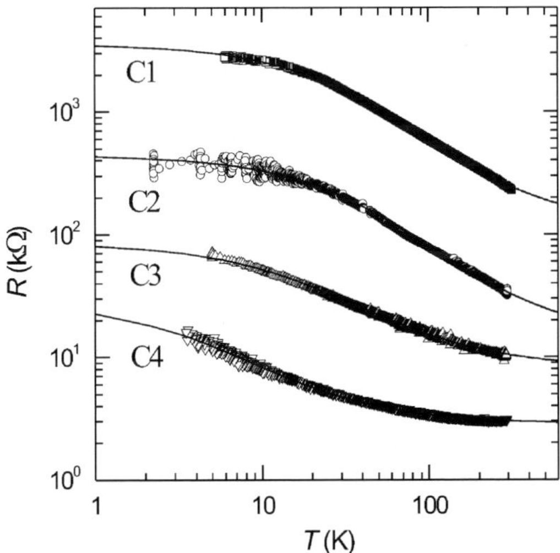

Fig. 12.4 Double logarithmic plot of the resistances versus temperature for four high-resistance nanowire devices, as determined from the 2-probe measurement configuration. The symbols are the experimental data and the solid curves are the theoretical fits to Eq. (12.1)

Table 12.1 Values of relevant parameters for four high-resistance nanowire devices, as determined from the 2-probe measurement configuration. For each device, the nanowire resistance R_s(300 K) was estimated from the 4-probe measurement configuration, while the junction area A was determined from the SEM image (see text) .

	R (300 K) (kΩ)	R_s (300 K) (kΩ)	R_0 (kΩ)	T_1 (K)	T_0 (K)	A (μm^2)	w (nm)	V_0 (meV)
C1	241	0.67	112	316	91	0.023	6.6	4.2
C2	33	0.33	13.6	363	104	0.022	6.4	4.5
C3	10.3	0.31	8.08	87	37	0.040	7.0	1.7
C4	3	0.33	2.82	20	8.6	0.028	8.7	1.1

increases rapidly as the temperature decreases below room temperature, making this approximation greatly satisfied.

It should be noted that, in the fluctuation-induced tunnelling model, since the two conducting regions remain large in size, the charging energy E_c needed to transfer an electron from one conducting region to the other is completely negligible, i.e., $E_c \ll k_B T$. This situation is very different from that in the case of the Coulomb blockade which involves fine metal grains or quantum dots, where the charging energy E_c ($\gg k_B T$) rather than the thermal voltage fluctuations plays the crucial roles in governing the electronic transport properties. In the present work, the volume of our "long" nanowire is relatively large as compared to the sizes of the fine metal grains (for example, $\sim 10^3$ nm^3) used in the Coulomb blockade studies (Tinkham et al., 2000). The typical volume of our nanowires is ~ 100 nm $\times$ 100 nm $\times$ 3 μm. Thus, our nanowire can be envisioned as a large conducting region separated by an insulating layer from another large conducting region (the submicron Cr/Au electrode) with a junction area A ($\sim 100 \times 500$ nm^2). This size of A is already small enough to render the aforementioned thermal voltage fluctuations important while large enough to make the Coulomb blockade phenomenon irrelevant. In effect, our nanowire devices in the 2-probe configuration mimic two similar tunnel junctions in series in the context of the fluctuation-induced tunnelling model.

Previously, the fluctuation-induced tunnelling model has been successfully applied to explain the temperature behaviour of the resistances in, among others, carbon polyvinylchloride composites (Sheng et al., 1978; Sichel et al., 1978, 1981), polymer composites (Paschen et al., 1995), and tin-doped indium oxide thin films (Ederth et al., 2003). In those "macroscopic" composite samples, a very large number of tunnel junctions with barely known junction parameters were often involved. On the contrary, the situation is greatly simplified and straightforward in our case, because in the 2-probe measurement configuration we deal with only two electronic contacts characterized by similar junction parameters, as discussed. Moreover, our junction area A is known. Interestingly, our experimental values of w and V_0 listed in Table 12.1 are on the same orders of magnitude to those obtained in carbon polyvinylchloride composites (Sheng et al., 1978; Sichel et al., 1978, 1981). This coincidence may be due to the fact that the sizes of our nanowires are approximately the same to the mean size of the conducting chains found in those composites.

Finally, it is worth noting that, if in our electronic contacts, the effective junction area is somewhat reduced from the maximum possible area A defined above, our values of the barrier width w would be slightly decreased, while values of the barrier height V_0 slightly increased, from those listed in Table 12.1.

12.4 Conclusion

We have measured the temperature dependent resistivities of individual single-crystalline RuO_2 nanowires, using both the 4-probe and the 2-probe methods. In the 4-probe measurements, the resistance reveals metallic behaviour, as expected. By employing the 2-probe method, we have quantitatively characterized the electronic contact resistances formed at the interfaces between the submicron electrodes and the nanowires. We found that the temperature behaviour of the high-resistance contacts can be well attributed to the thermally fluctuation-induced tunnelling conduction. The junction parameters such as the barrier width and height have been determined. This work demonstrates that, under certain conditions, the electronic contacts between interconnects and metal nanodevices can be quantitatively modelled. In further studies, the junction parameters w, V_0, and A may be tailored by adjusting the diameter and surface condition of the nanowire, the electrode material, and the width of the electrodes, etc. Finally, we should point out that the electronic contact resistance is a quantity very sensitive to the interface condition and, in different classes of nanowires and electrodes, the temperature behaviour may be governed by a mechanism different from the fluctuation-induced tunnelling conduction (Lin et al., 2007).

Acknowledgments The authors are grateful to T. C. Lee, S. P. Chiu, S. J. Hong, Y. L. Liu and J. J. Huang for their experimental help, and to Ping Sheng for valuable discussion. This work was supported by the Taiwan National Science Council through Grant Nos. NSC 93-2120-M-009-009 and NSC 94-2120-M-009-010, and by the MOE ATU Program.

References

Ederth, J., Johnsson, P., Niklasson, G. A., Hoel, A., Hultåker, A., Heszler, P., Granqvist, C. G., van Doorn, A. R., Jongerius, M. J. and Burgard, D., 2003, Electrical and optical properties of thin films consisting of tin-doped indium oxide nanoparticles. *Physical Review B*, **68**, 155410.

Gu, G., Burghard, M., Kim, G. T., Düsberg, G. S., Chiu, P. W., Krstic, V., Roth, S. and Han, W. Q., 2001, Growth and electrical transport of germanium nanowires. *Journal of Applied Physics*, **90**, 5747.

Ip, K., Heo, Y. W., Baik, K. H., Norton, D. P., Pearton, S. J., Kim, S., LaRoche, J. R. and Ren, F., 2004, Temperature-dependent characteristics of Pt Schottky contacts on n-type ZnO. *Applied Physics Letters*, **84**, 2835.

have so large specific surface area that it is easy to be doped with impurities of some elements and be introduced into structural defects (Zhang and Zhang, 2002; Jeong et al., 2004). The III-V group nitrides recently become very attractive in applications for light emitters, laser diodes, and optoelectronic devices in the ultraviolet and visible spectral range (Nakamura, 1998), because the wurtzite nitrides can form a continuous alloy system with adjustable direct band gaps from 6.2 to 1.9 eV. Hence, ultimate nanoscale optoelectronic devices for a wide range of wavelengths with many advanced features such as high thermal conductivity and superior stability may be realized by using these 1D nanostructures. To date, intensive studies have been devoted to the synthesis, properties and prototypes of 1D III-V group nitrides (Haber et al., 1997, 1998; Liu et al., 2001, 2005b; Zhang et al., 2001; Yin et al., 2005a, b; Tondare et al., 2002; Tang et al., 2005; Wu et al., 2003, 2004; Zhao et al., 2005; Shi et al., 2005). The nitrides series possess wide-range band gaps, are therefore ideal for the development of solid-state white light emitting devices (Vurgaftman et al., 2001). AlN is the only wurtzite compound that has been predicted to have a negative crystal field splitting at the top of valence band which can lead to unusual optical properties (Suzuki et al., 1995; Wei and Zunger, 1996; Chen et al., 1996). An AlN PIN (p-type/intrinsic/n-type) homojunction LED with an emission of the shortest wavelength of 210 nm has been successfully developed (Tanniyasu et al., 2006). In a word, IIIA group nitride semiconductors are potential materials for ultraviolet LEDs and laser diodes, but suffer from difficulties in controlling electrical conduction to achieve exciton-related light-emitting devices as well as replacing gas light sources with solid-state light sources.

With the rapid development of microelectronics, electronic packaging materials have become vital to the success of advanced designs. AlN and BN are expected to be excellent reinforcements in composites for electronic packaging or structural materials because of their good mechanical properties, high thermal conductivity, low thermal expansion (CTE) of AlN ($\sim$3.3 $\times$ 10^{-6}/K) (Geiger and Jackson, 1989) is close to that of Si and GaAs, and low density. (Bradshaw and Spicer, 1999; Huang and Li, 1994) Moreover, with high chemical and thermal stability, AlN and BN do not react with Al, thereby avoiding the formation of brittle reaction products on the reinforcement/matrix interface. In fact, AlN micro- or nano-particles have been applied as reinforcement for metal and polyamide matrix composites because of the above characteristics. (Huang and Li, 1994; Lai and Chung, 1994; Inoue et al., 1993; Chedru et al., 2001; Vicens et al., 2002; Zhang et al., 2003; Chen and Gonsalves, 1997; Zhang et al., 2002; Wu et al., 2003) AlN nanowires, compared with conventional powder-form AlN, are predicted to exhibit better mechanical and thermal properties (Haber et al., 1998; Zhang et al., 2001) and better reinforcing performance due to their high aspect ratio and near-perfect crystal structure (Haber et al., 1998; Zhang et al., 2001; Liu et al., 2001, 2005). Combining Al with 1D AlN nanostructures may create novel nanocomposites with high strength and thermal conductivity, as well as low density and CTE. Although nitride nanowires or nanotubes were synthesized in the past years, (Haber et al., 1998; Zhang et al., 2001; Liu et al., 2001, 2005b; Yin et al., 2005a, b; Tondare et al., 2002; Tang et al., 2005; Wu et al., 2003, 2004; Zhao et al., 2005; Shi et al., 2005) their reinforcement behaviors in composites have not been investigated yet.

It is known that many unique properties of materials depend on their structural characteristics, which has stimulated the on-going interest in synthesizing various nanostructures with different morphologies. Considerable effort has been made on the synthesis of 1D nitride nanostructures as well. 1D nitride nanostructures such as nanotubes, nanowires, and nanorods are successfully synthesized by several routes, including chloride-assisted growth (Haber et al., 1998; Liu et al., 2005), carbon nanotube-confined reaction (Zhang et al., 2001, Liu et al., 2001), arc-discharge (Tondare et al., 2002; Tang et al., 2005c), direct nitridation (Wu et al., 2003a, b; Zhao et al., 2005), and vapor transport and condensation (VTC) (Yin et al., 2005). Despite of the recent progress in preparing 1D nitride nanostructures such as nanotubes (Yin et al., 2005b; Tondare et al., 2002), nanowires (Haber et al., 1998; Zhang et al., 2001; Liu et al., 2001, 2005), nanobelts (Tang et al., 2006), and comb-like morphologies (Yin et al., 2005a) by different methods, the controllable synthesis or regulation of these nanostructures is still a remained challenge, and the growth mechanism of 1D nitrides is not yet clear.

In this study, high purity nitride nanostructures, including BN nanotubes and nanofibers, AlN nanorods, nanotips, nanobelts and nanofibers are synthesized (Tang et al., 2005, 2006a, 2007; Liang and Che, 1993). Especially, based on the understanding of the growth mechanism, single-crystalline AlN mushroom-like nanorod array, Eiffel-tower-shape nanotip array, honeycomb-like network of vertically aligned nanoplatelets and flowerlike Si-doped AlN nanoneedle array have been intentionally synthesized. Moreover, the field emission properties, optical properties, mechanical properties and thermo-physical properties of these nanostructures and their composites are investigated.

13.2 Experimental

Most of the 1D nitride nanostructures were synthesized using a chemical vapor transport and condensation (CVTC) method. The synthesis was carried out in a conventional electric resistance furnace with a horizontal alumina tube. In a typical process, 1.5 g Al (99.5%) nanoparticles ($\sim$80 nm in size) was used as raw material and placed in a ceramic boat which was set at the center of the alumina tube. A Si wafer was put at the downstream end of the ceramic boat with a separation of 1.0 cm from raw reactants in order to collect the grown product. Argon gas was kept flowing into the alumina tube at a rate of 200 sccm to minimize any side reactions before the furnace was heated to the reaction temperature. Some metals or oxides were selectively used to be raw materials or catalyst. The Al source was maintained at certain temperatures and NH_3 was introduced at a rate of 80 sccm for 1 hour. Then the furnace was cooled down to room temperature under flowing NH_3, a layer of light-gray product was found on the Si wafer, the ceramic boat and the inner wall of downstream-end tube. The product was collected from the Si wafer for later characterization.

A vacuum apparatus that can produce a plasma electric arc within a big reactor was employed for the synthesis of AlN nanorod arrays. The stainless steel reactor chamber has a diameter of 600 mm and height of 400 mm. Commercial Al (100 $\times$

$100 \times 30\,mm^3$) with a purity of 99.98% was employed as both the raw material and the anode of the vacuum arc discharge apparatus, and it was placed under the tip of the tungsten cathode (the diameter was 10 mm) with a distance about 5–10 mm.

The phase purity and crystal structure of the products were detected by X-ray diffraction (XRD, Regaku D/max-2400). A scanning electron microscopy (SEM, Philips XL30) was used to observe the morphology of the as-prepared samples. Further detailed structural characterization and elemental composition analysis were performed on a high resolution transmission electron microscopy (HRTEM, JEOL JSM-2010) equipped with an energy dispersive spectroscopy (EDS, Oxford). Photoluminescence (PL) spectrum of the AlN nanostructures was measured using a He-Cd laser (325 nm) as the excitation source at room temperature.

13.3 Results and Discussion

13.3.1 Synthesis of 1D Nitrides

13.3.1.1 Platelet BN nanowires

The platelet BN nanowires were synthesized by using amorphous boron powder and boron oxide powder as boron source, NH_3 as nitrogen source, nickelocene as catalyst precursor, and FeS as a growth promoter (Tang et al., 2006a). The raw chemicals (except for nickelocene) were evenly mixed and put into an h-BN boat covered with a graphite substrate, and then placed at the centre of the furnace, while nickelocene was placed upstream of the reactor. After fluxing with Ar for half an hour, NH_3 and Ar mixture gas at a ratio of 1:8 was introduced. The temperature was increased to 1350° within 35 minutes and kept unchanged for 1 hour. Then, the furnace temperature was allowed to lower down to room temperature gradually. The as-prepared products are white cotton-like substance deposited on the graphite substrate. By tuning the growth time, gas flow rate, and the deposition position, the platelet BN nanowires can be selectively obtained (Fig. 13.1). We can see that the nanowires are not uniform in diameter along their axis direction, and they are actually composed of numerous series-wound cup-like sections (Fig. 13.1a, b). The lengths of the "cups" range from tens of nanometers to 500 nm. The cone angles of the "cups" are similar within one nanowire, while they may differ from each other for different nanowires. HRTEM observation (Fig. 13.1c) shows that the cup-like section is single-crystalline and their (0002) planes are perpendicular to the axis direction with an interplanar distance of 0.34 nm. The chemical composition of the nanowires was analyzed by using electron energy loss spectroscopy (EELS). A typical EELS spectrum is shown in Fig. 13.1d. Clear K edges of boron and nitrogen are shown at around 188 and 401 eV, respectively. No noticeable C-K edges (~284eV) can be detected.

13.3.1.2 AlN Nanorod Array Controlled by Temperature Gradient

A vacuum apparatus that can produce a plasma electric arc was employed for the synthesis of AlN nanorod arrays (Fig. 13.2, Tang et al., 2005c). Commercial Al

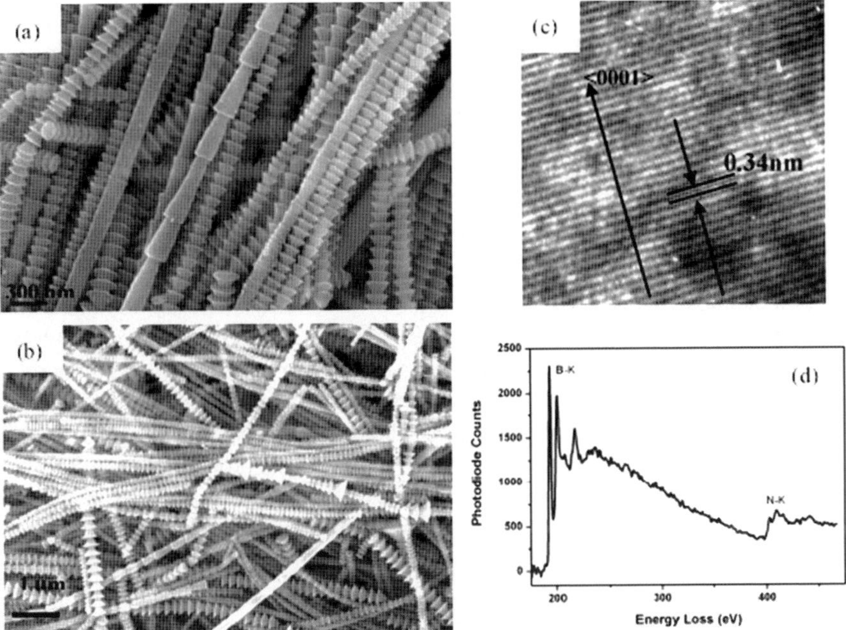

Fig. 13.1 SEM (a, b), TEM (c) images, and an electron energy loss spectrum (d) of the as-prepared platelet BN nanowires

$(100 \times 100 \times 30\,\text{mm}^3)$ with a purity of 99.98% was employed as both the raw material and the anode of the vacuum arc discharge apparatus, and it was placed under the tip of a tungsten cathode (the diameter was 10 mm) with a distance about 5–10 mm. The discharge current was kept at 150 A throughout the reaction. After the arc discharge proceeded for 30 minutes, a layer of white products was obtained around the anode Al.

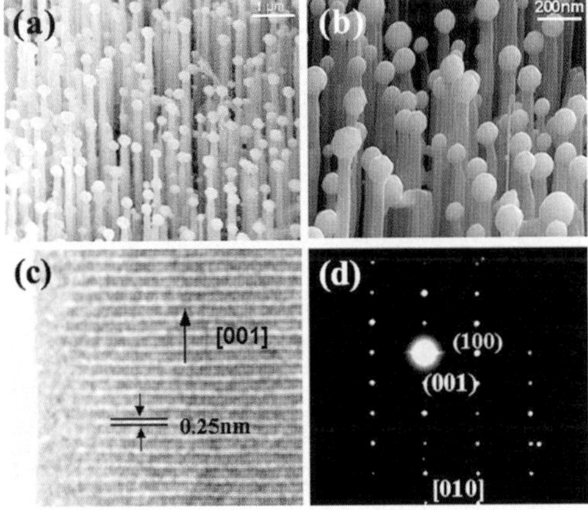

Fig. 13.2 The SEM images of AlN nanorod array: (a) tilted top-view, and (b) high magnification. High-resolution TEM image of AlN nanorod (c) and the corresponding ED pattern (d)

The essence of 1D nanostructure formation is about crystallization which involves two fundamental steps: nucleation and growth (13.1). The nucleation of a typical vapor-liquid-solid (VLS) process are mainly induced and dictated by catalyst liquid droplets. Without using catalyst, the growth mechanism of the single-crystalline mushroom-like AlN nanorod arrays should be a special VLS process. Figure 13.3A schematically illustrates three major steps involved in the growth process of the AlN nanorod. Based on the Al-N binary phase diagram (Fig. 13.3B) (Tang et al., 2005b), Al and N form AlN at very low nitrogen partial pressure when the temperature is raised above the eutectic point (660.45°). The reaction can be expressed as: $N(g) + Al(l) \rightarrow AlN(s)$. Furthermore, the higher the reacting temperature, the quicker the reaction. Due to the high current densities and plasma temperatures involved, target material melts and atomic flux protuberated from the molten surface reacts with ionized nitrogen to form AlN nuclei by a homogeneous nucleation process (step I) (Haber et al., 1998). This mechanism is supported by the SEM observation (Fig. 13.3C) of the product obtained on the initial stage of synthesis process, which clearly records the nucleation of AlN nanorods. Subsequently (step II), the location of liquid isotherm of AlN is gradually raised to form an extending temperature gradient field as the tungsten cathode was moved upward, and the AlN nuclei served as active sites for Al cluster adsorption

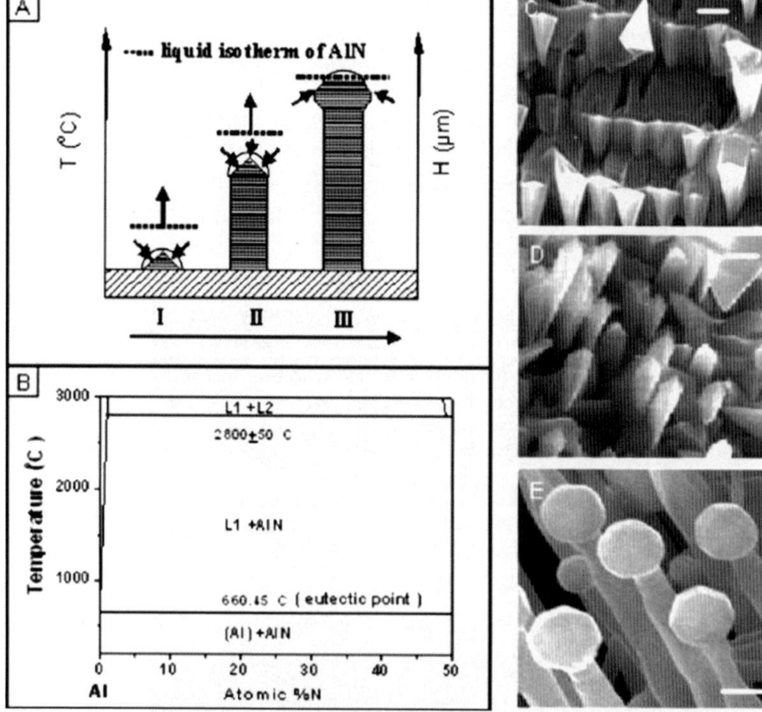

Fig. 13.3 The growth mechanism of the AlN nanorod arrays: (A) schematic illustration three major steps in the growth process, (B) the Al-N binary phase diagram, (C) the SEM observation in step I, (D) the SEM observation in step II, (E) the SEM observation in step III, the scale bar in (C)–(E) is 200 nm

would preferentially grow along the temperature gradient which is also [001] of the nanorod crystal. In this process, the temperature gradient plays an important role to induce the growth of AlN nanorod. A SEM image of the product obtained on the middle stage of synthesis process is shown in Fig. 13.3D, which obviously proves the growth process of AlN nanorod. As the cathode was kept at a certain height, the speed of the growth along the temperature gradient would slow down and form the droplet of [Al(l)+AlN(s)] on the top of AlN nanorod which would react subsequently with the nitrogen to form a polyhedral AlN ball (Fig. 13.3E). The reason for the formation of the polyhedrical end of the nanorod is likely due to reducing the surface area to decrease the surface energy according to the Wulff's theorem. In a word, unlike the typical catalyst induced VLS growth mechanism, temperature gradient is one of the key factors in the nucleation and growth process.

13.3.1.3 Oxide-assisted Growth of AlN Nanotip Array Controlled by the Amount of Reactants

AlN nanotip array has also been synthesized by a CVTC method (Tang et al., 2005a). Some typical SEM images of the as-synthesized product are shown in Fig. 13.4a, b and c. It is very interesting that the morphology of the product shows an Eiffel-tower-shape nanostructure, as shown in Fig. 13.4a. Most of the deposited AlN nanotips are oriented vertically with respect to the Si substrate, while some tips with different directions are occasionally observed (Fig. 13.4b). Independent of their size, AlN nanotips are always composed of two parts, a submicron-sized base

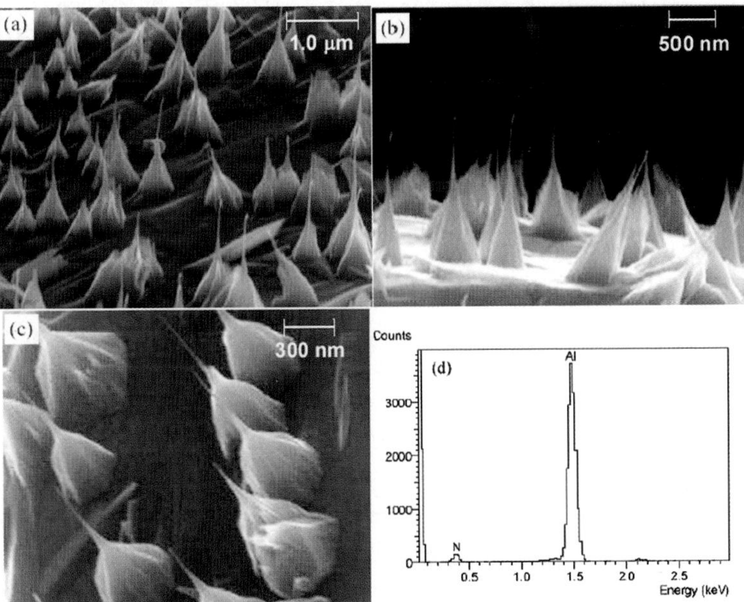

Fig. 13.4 SEM images of AlN nanotip array: (a) tilted top view, (b) side view and (c) high magnification, and (d) EDS spectrum

and a sharp tip. As shown in Fig. 13.4c, the base is a polygonal pyramid with a few hundred nanometers in size, and the sharp tip is several to tens of nanometers in diameter and $\sim$1.0μm in length. The EDS spectrum (Fig. 13.4d) indicates that these nanostructures only consist of Al and N elements.

A representative transmission electron microscopy (TEM) image of an individual nanotip is shown in Fig. 13.5a, which was taken along one of its side surface perpendicular to the electron beam. Selected area electron diffraction (SAED) patterns were recorded during TEM observations. SAED pattern (Fig. 13.5b) taken from the region marked by a circle in Fig. 13.5a can be indexed as that of single-crystalline wurtzite AlN along the [010] zone axis. Figure 13.5c presents the corresponding HRTEM image which was taken near the edge and along the length of this tip. The spacing between any two adjacent lattice fringes (0.25 nm) corresponds to that of the (002) lattice planes of the hexagonal AlN. The axis of the tip is also perpendicular to the lattice plane (002), that is to say, the axis direction of AlN nanotips is generally along [001]. Both the HRTEM and SAED results demonstrate that the as-grown nanotips are single-crystalline growing along [001] direction of AlN hexagonal wurtzite structure.

Experiments were carried out to study the role of each reactant played in growth and to explore the growth mechanism. To verify the formation mechanism of the Eiffel-tower-shape AlN nanotips, and to see whether the starting material without Fe_2O_3 is suitable for synthesizing this nanostructure, only Al nanoparticle was used as Al source for the reaction under similar experimental conditions; but no tips similar to those obtained with Fe_2O_3 were formed. We also used the mixture of Al and Al_2O_3 nanoparticles as raw reactant under the same condition and did not observe any nanostructures as well, only leading to the deposition of amorphous particles of AlN (several microns in size) on Si substrate. It can therefore be concluded that

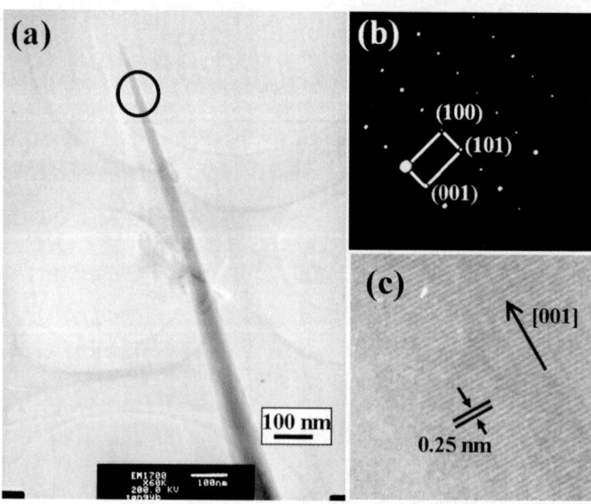

Fig. 13.5 TEM image (a), SAED pattern (b) and the corresponding HRTEM image (c) of a single AlN nanotip

Fe_2O_3 plays a key role in this synthesis process. It is speculated that the starting material will undergo the following reactions:

$$2Al(l) + 2NH_3(g) \rightarrow 2AlN(s) + 3H_2(g) \, , \, \Delta G = -401.72 \, kJ/mol \qquad (13.1)$$

where ΔG is the change of Gibbs free energy of the reaction at 1000 K. This reaction is highly exothermic, which may result in substantial local heating, therefore promoting the following reactions (Zhang et al., 2001; Liu et al., 2001). Subsequently, Al_2O vapors are obtained from the chemical reaction:

$$6Al(l) + Fe_2O_3(s) \rightarrow 3Al_2O(g) + 2Fe(s), \, \Delta G = -106.69 \, kJ/mol \qquad (13.2)$$

In the previous synthesis methods (Zhang et al., 2001; Liu et al., 2001), the Al_2O vapor is difficult to obtain at low temperature because the Al_2O_3 is hard to reduce by Al or carbon due to its very low Gibbs free energy (for example, $G(Al_2O_3)$ is $-1777 \, kJ/mol$ at 1000 K) (Liang et al., 1993). In this case, we used Fe_2O_3 instead of Al_2O_3 as oxidizer and it can significantly facilitate reaction (13.2) to proceed. The Al_2O vapor will reacts with NH_3 to generate AlN species:

$$Al_2O(g) + 2NH_3(g) \rightarrow 2AlN(s) + H_2O(g) + 2H_2(g),$$
$$\Delta G = -518.22 \, kJ/mol \qquad (13.3)$$

Reaction (13.3) is a vapor-vapor reaction between Al_2O and NH_3. The formed AlN species can be readily transported to the deposition zone by the carrier gas and grow into AlN crystals.

It is well known that the nucleation of a typical vapor-liquid-solid (VLS) process are mainly induced and dictated by catalyst liquid droplets. In our experiment, the growth of the nanotips should be governed by a vapor-solid epitaxy process. From the morphology of the tips (Fig. 13.4), it is believed that the nanotips are formed by two steps, as schematically illustrated in Fig. 13.6a. Step I: growth of the submicron-sized base initially on the Si substrate and step II: growth of nanoneedle on the top of the base. In the initial stage, the abundant reactants provide sufficient AlN vapor, therefore, the growth of AlN is so rapid that relatively big bases are formed. With the deposition going on, the concentration of AlN atoms reduces gradually due to limited supply from the source, so the quantity of AlN atoms at the VS interface becomes depleted, which leads to a gradual reduction of the area of VS interface along the growth direction, and brings on the growth of AlN nanoneedles. The SEM observation (Fig. 13.6b) of the product obtained at the middle stage of the synthesis process (50 minutes) clearly records the growth process of this nanostructure.

13.3.1.4 AlN Honeycomb-like Network Controlled by the Surface Conditions of Substrate

AlN nanoplatelets were synthesized by heating the mixture of Al and $AlCl_3$ powders in the flow of N_2 (Tang et al., 2006b). Briefly, a Si wafer ($2 \times 3 \, cm$) was used as

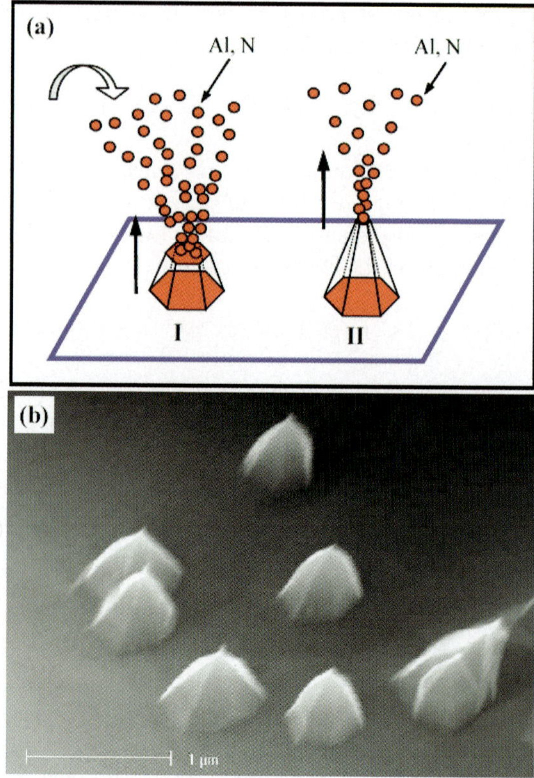

Fig. 13.6 The schematic illustration and SEM observation of controlled growth of AlN nanotip array

the substrate, which was etched using 2% hydrofluoric acid for about 5 minutes and ultrasonically cleaned by acetone and de-ionized water several times to remove silicon oxide on its surface. The Si substrate was placed in a ceramic boat and deposed at downstream of the gas flow with a distance of $\sim$15 cm from the reactants. The Ar gas was substituted by N_2 (50 sccm) as the furnace was heated to 1200 °C at a rate of 30 °C/min. The reaction was maintained for 1 hour at a pressure of 3 Torr. After the furnace was naturally cooled down to room temperature under the flow of Ar, a layer of white product was observed on the surface of the Si substrate. However, when the non-etched Si wafer was used as the substrate and other conditions were kept the same, we did not get similar results.

The AlN nanoplatelets synthesized on Si substrate are shown in Fig. 13.7. The nanoplatelets vertically grow on the Si substrate and connect with each other to form a honeycomb-like network. The thickness of the single-crystalline nanoplatelets varies from 10 to 100 nm, and their height varies from several hundred nanometers to several microns (Fig. 13.8).

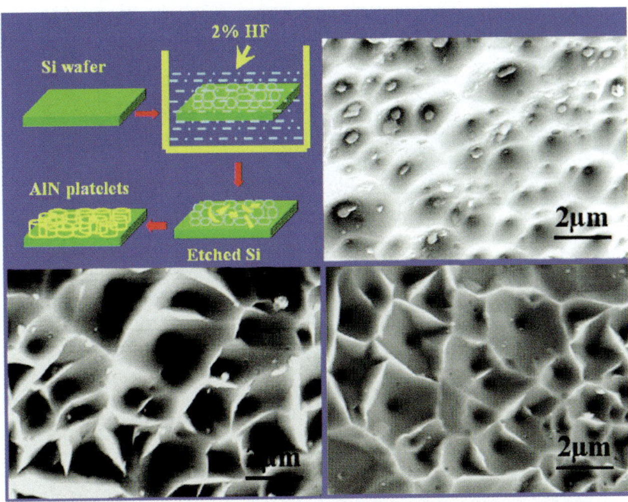

Fig. 13.7 The schematic illustration and SEM observations of controlled growth of AlN nanoplatelets

13.3.1.5 Catalyst-seeded Synthesis of Flowerlike Si-doped AlN Nanoneedle Array

Si wafers (1×2 cm), on which cobalt (Co) catalyst particles (1–$3\,\mu$m in diameter, 99.9%) were evenly dispersed (Fig. 13.9a], were employed as substrate for the growth of Si-doped AlN nanostructures (Tang et al., 2006c). In a typical process, 1 g Pluronic P-123 triblock copolymer ordered from Aldrich was dissolved in 30 ml ethanol, and 0.10 g Co particles were added to this solution and then stirred for 30 minutes to form an emulsion. The catalyzed Si substrates were prepared by coating the emulsion solution containing Co particles on Si wafers. Before being loaded into the furnace, the coated Si substrates were annealed at 500 °C in hydrogen atmosphere for 20 minutes to remove the polymer. $AlCl_3$ (99.9%) powders were placed in an alumina boat and the catalyzed Si substrate was put downward on the top of the source boat at the center of a horizontal tube furnace. After that, the furnace was purged with argon gas which was then substituted by NH_3/H_2 (60 ml/min, 2:1 in volume) as the furnace was heated to 900 °C. About 20 ml liquid $SiCl_4$ (99.9%) was gradually fed into the tube by an injector. A white-color thin film formed on the substrate after about 30 minutes reaction.

A typical SEM image of the AlN nanostructures synthesized at 900 °C is shown in Fig. 13.9b, displaying the sample is an AlN nanoneedle array with a flower-like morphology, which were obtained in MoS_2, GaP, and MgO, etc. (Liu et al., 2005a; Feng et al., 2006; He et al., 2006). As an example, Fig. 13.9c presents a complete AlN micro-flower, from which we can see that numerous AlN nanoneedles grow radially from a seeded Co catalyst particle. These nanoneedles are not absolutely straight, and they are several micrometers long and gradually taper along the growth

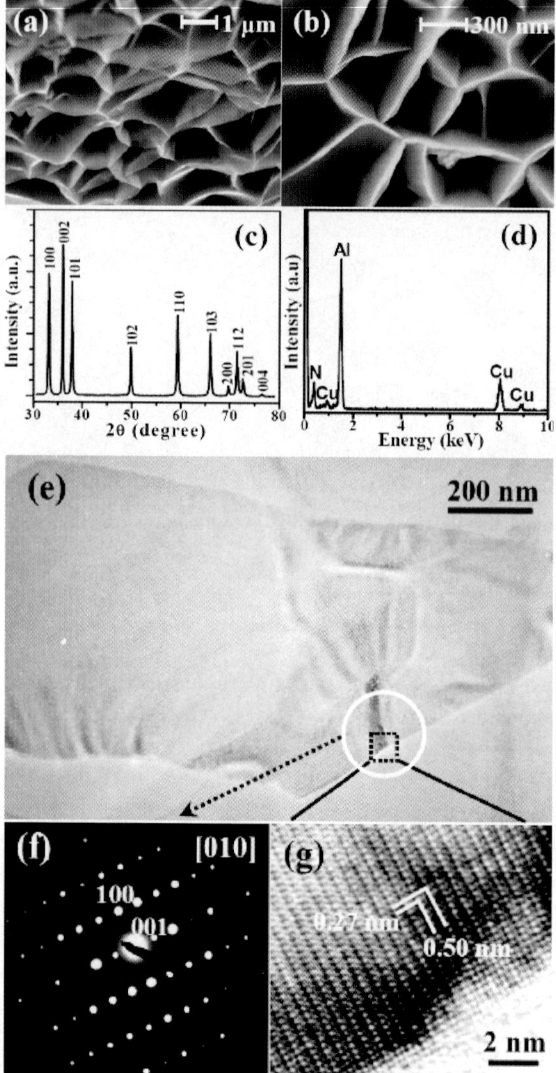

Fig. 13.8 The SEM images of AlN nanoplatelets: (a) tilted top-view and (b) high magnification. XRD diffraction spectrum (c) and EDS spectrum (d) of AlN nanoplatelets. TEM image (e), SAED pattern (f) and the magnified HRTEM image (g) of a single AlN nanoplatelet

direction (Fig. 13.9d). The base diameters of the nanoneedles are in the range of 50–150 nm, and the tips are 5–30 nm in diameter. By varying the concentration of the seeding emulsion, the area density of the nanoneedle array on the substrate is able to be adjusted.

Figure 13.10a is a TEM image of an AlN nanoneedle. The selected area electron diffraction (SAED) pattern (Fig. 13.10b) reveals that the nanoneedle

Fig. 13.9 (a) The Si substrate covered with Co catalyst particles, and (b)–(d) SEM images of flower-like AlN nanoneedle array synthesized at 900 °C

is single-crystalline and grows along the [001] direction. Figure 13.10c is the magnified HRTEM image taken from the tip of this nanoneedle. The distance between adjacent lattice planes perpendicular to the growth direction is 0.50 nm, corresponding to the spacing of (001) planes of hexagonal AlN. Figure 13.11 displays a TEM image of an AlN nanoneedle, and its corresponding Al, N, and Si elemental mappings. It reveals that Si element is uniformly distributed in the AlN nanoneedle.

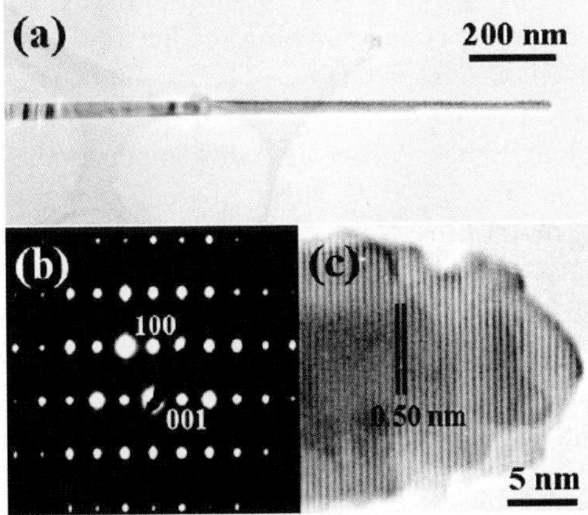

Fig. 13.10 TEM image (a), SAED pattern (b), the magnified HRTEM image (c)

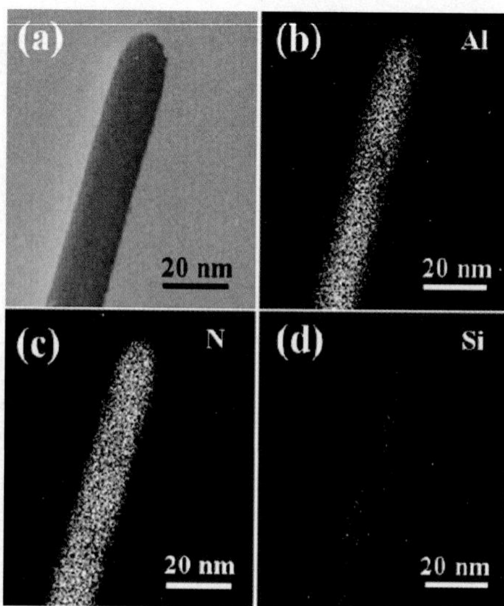

Fig. 13.11 TEM image of an AlN nanoneedle (a), and its corresponding Al (b), N (c), and Si (d) elemental mappings

The catalytic Co particles are believed to be responsible for the formation of flower-like AlN nanoneedles since no similar nanostructures were formed on the Si substrate without covering Co particles under similar experimental conditions. The growth of the Si-doped AlN nanoneedles should be a base-model VLS process which is similar to the previous reports where Ni films were used as catalyst for synthesizing 1D AlN nanostructures (Liu et al., 2005b; Xu et al., 2004). At the reaction temperature, Al, N, and Si vapors are generated by thermal decomposition of $AlCl_3$, NH_3, and $SiCl_4$, and they deposit on the surface of the Co particles to form eutectic liquid droplets. Solid AlN nuclei will precipitate from alloy droplets that become supersaturated with Al and N supplied from the gas phase. Meanwhile, Si doping is achieved by the process of Si substituting Al atom in AlN crystal. Continuous dissolving of Al, N and Si atoms into the liquid droplets maintains the growth of the Si-doped AlN nanostructures. The growth rate of the AlN nanostructures in the axial direction far exceeds that in the radial direction, resulting in a needle-like morphology, that is, AlN nanoneedles. In this study, when the synthesis temperature was higher than 1000 °C, larger AlN nanocones could be obtained as shown in previous reports (Liu et al., 2005b; Zhao et al., 2005), which probably originates from the fact that the radial growth becomes more obvious at a higher temperature. Nevertheless, the detailed growth mechanism still needs further study.

13.3.1.6 AlN Nanobelt Array

The synthesis of AlN nanaobelt arrays is a CVTC process. Al and Fe_2O_3 nanoparticles (~50 nm in size) were fully mixed in a weight ratio of 1:1. When the temperature of the furnace reached 900 °C, the Ar gas was substituted by 80 sccm Ar/NH_3 (3:1 in volume) for 1 hour. The Si substrate was covered with a layer of fluffy thin film after the reaction.

Figure 13.12a shows the typical SEM image of an as-grown sample deposited on the Si substrate. It can be seen that high-purity vertically aligned 1D nanostructures distribute relatively uniformly on the surface of the Si substrate. The length of this 1D structure is in the range of 1–3 µm. Some nanostructures are bent along their length without being broken, which indicates that they are flexible and probably belt-like in nature. TEM observations further confirm the belt-like morphology of these 1D nanostructures. The uniform contrast and electron beam transparent characteristics demonstrate that the formed structures are nanobelts (Fig. 13.12b). The ripple-like contrast in the TEM image is due to the bending of the nanobelts. The width of the belts is in the range of 20–150 nm and the thickness is in the range of 10–30 nm. The ratio of width to thickness of the belts ranges from 2 to 5, as estimated by TEM observations.

Figure 13.12c presents a TEM image of an individual nanobelt, which was taken along one of its side surfaces perpendicular to the electron beam and used for SAED and HRTEM observations. Its corresponding SAED is displayed in Fig. 13.12d. From the SAED pattern, the crystal structure of this belt is identified to be hexagonal wurtzite AlN phase. It also indicates that the belt is single crystalline and enclosed by $\pm(100)$ and $\pm(010)$ crystallographic facets. The HRTEM image taken near the edge of this nanobelt is shown in Fig. 13.12e. The adjacent lattice spacing in longitudinal direction is about 0.25 nm, corresponding to that of the (002) planes of wurtzite AlN, which further confirms that the nanobelt is wurtzite AlN phase. Meanwhile, it shows that the nanobelt grows along [001] direction, which is the frequent growth orientation for hexagonal wurtzite-structure materials. In addition, a $\sim$2 nm thick amorphous coating is clearly seen along the surface of the belt.

13.3.1.7 AlN Nanofibers

The mixture of Al and Fe_2O_3 nanoparticles (less than 100 nm in size) with a weight ratio of 5:2 was used as raw material and was placed in a ceramic boat which was set at the center of the alumina tube. The furnace was rapidly heated to and maintained at 800 °C for 1.5 hours under a flow of NH_3/Ar (2:1 in volume) at a rate of 150 sccm. After the furnace was cooled down to room temperature, a layer of white wool-like product was found on the surface of the substrate and it was collected for later characterization (Tang et al., 2005b).

The morphology of the as-synthesized products was observed by scanning electron microscopy (SEM). The typical SEM images of the products are shown in Fig. 13.13. The Si substrate is covered by a large number of randomly distributed wire-like nanostructures (Fig. 13.13a), and most of these 1D nanostructures have uniform diameter over their entire length. Figure 13.13b shows that some nanofibers are bent to a 180 angle without being broken, which demonstrates that the nanofibers are very flexible. To further verify the geometrical morphology of the AlN nanostructures, a high magnification SEM image of the nanofibers is shown in Fig. 13.13c,

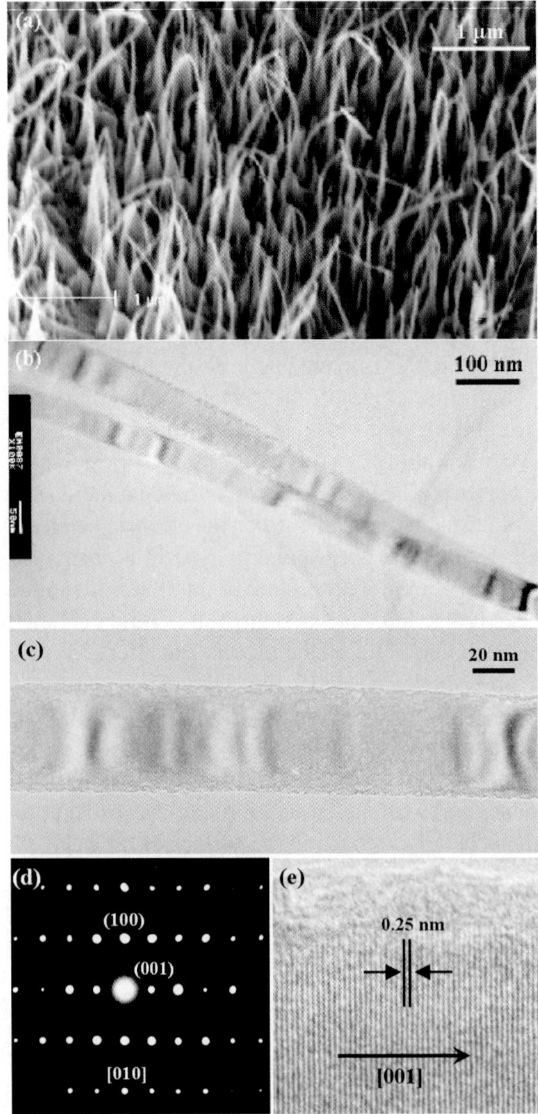

Fig. 13.12 (a) The typical SEM image of the AlN nanobelt array on the Si substrate. (b) Low-magnification TEM image of the AlN nanobelts. (c) a high-magnification TEM image, (d) SAED pattern, and (e) the corresponding HRTEM image of a single AlN nanobelt

which displays prism morphology. The nanofibers are several tens of micrometers in length and 10–200 nm in width.

Figure 13.14a shows the representative TEM image of the end of a nanofiber. It can be seen that this nanofiber has a rectangular cross-section shape and the ratio of width to thickness is in the range of 1–2, which is different from the previous reported nanobelts (Tang et al., 2006). No particles were observed at the tip of the

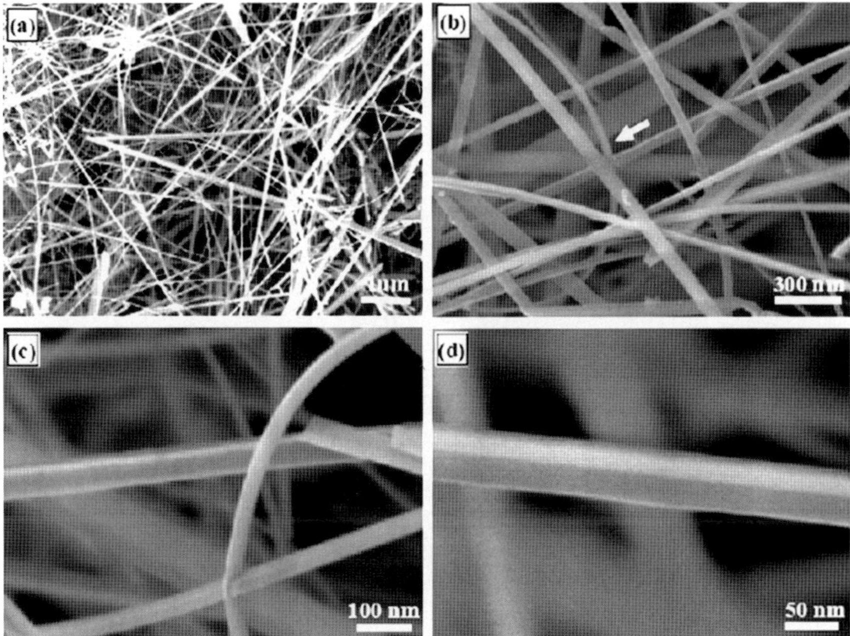

Fig. 13.13 SEM images of AlN nanofibers

nanofibers during the TEM observations. The typical EDS spectrum (Fig. 13.14b) shows that the nanofibers are composed of Al and N elements (the Cu signal comes from the copper grid) with an approximate atomic ratio of 1:1. The further detailed microstructures of the rectangular cross-section AlN nanofibers were studied by high resolution TEM (HRTEM) and selected area electron diffraction (SAED). TEM image of an individual nanofiber is shown in Fig. 13.14c, which was taken oriented with one of its side surfaces perpendicular to the electron beam. The corresponding SAED pattern (Fig. 13.14d) was recorded along the [010] zone axis the hexagonal wurtzite AlN. This indicates that the nanofiber is single-crystalline and it is enclosed by (100) and (010) crystal planes. Fig. 13.14e is the HRTEM image which was taken near the edge along the length of this nanofiber. The adjacent lattice distances in directions of parallel and perpendicular to the nanofiber axis are 0.27 and 0.50 nm, respectively, corresponding to the d-spacing of (100) and (001) crystal planes of hexagonal AlN. Meanwhile, the axis of the nanofiber is perpendicular to the lattice plane (001), which suggests the growth direction of the nanofiber is along [001] direction, which is the frequent growth orientation for hexagonal wurtzite structure materials. A 1–2 nm thick amorphous coating is clearly seen along the surface of the nanofiber. This thin layer is determined to be aluminum oxide (AlO_x) from the EDS spectrum. Amorphous oxide coating is often found in AlN 1D nanostructures (Zhang et al., 2001; Liu et al., 2001). Both the HRTEM and SAED observations reveal that the nanofibers are single-crystalline wurtzite AlN which grow along [001] direction.

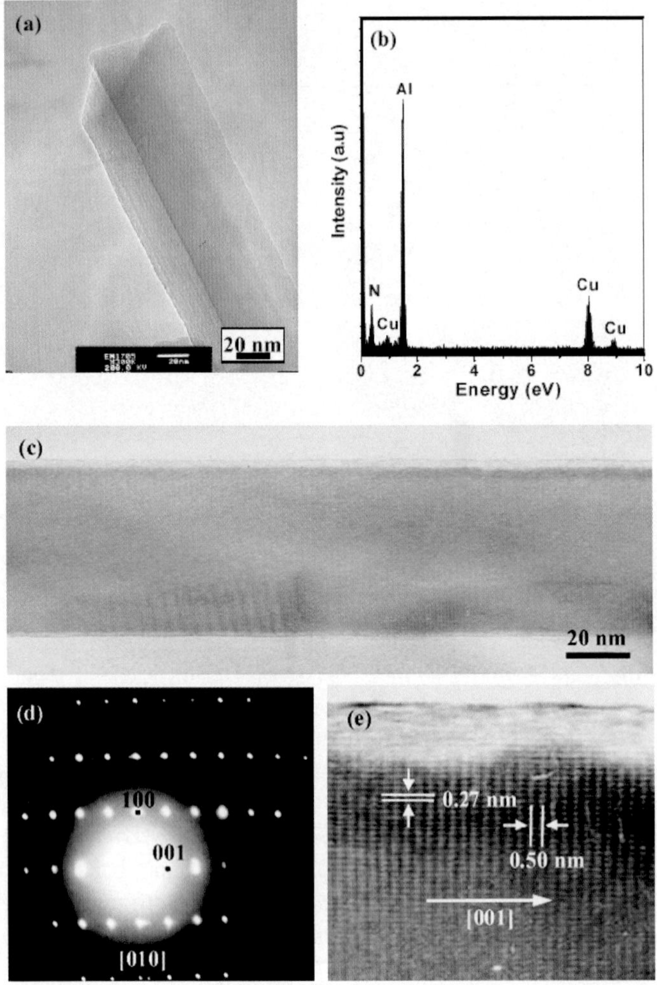

Fig. 13.14 (a) TEM image of the end of an AlN nanofiber. (b) The typical EELS spectrum of AlN nanofibers. (c) TEM image, (d) SAED pattern, and (e) the corresponding HRTEM image of an AlN nanofiber

13.3.2 Properties of 1D AlN

13.3.2.1 Field Emission of 1D AlN Arrays

The FE measurements for four kinds of 1D AlN arrays (Fig. 13.15) were carried out inside a ball-type chamber which was pumped down to 5.0×10^{-9} torr by an ultrahigh vacuum system. The anode was a cylinder-shaped platinum probe (1 mm in diameter) and the cathode was AlN nanotip array, with a distance of 100 μm. High voltage was supplied by a power source (Keithley 248) and the current under increasing applied voltage which varied with a step of 10 V was recorded by a sensi-

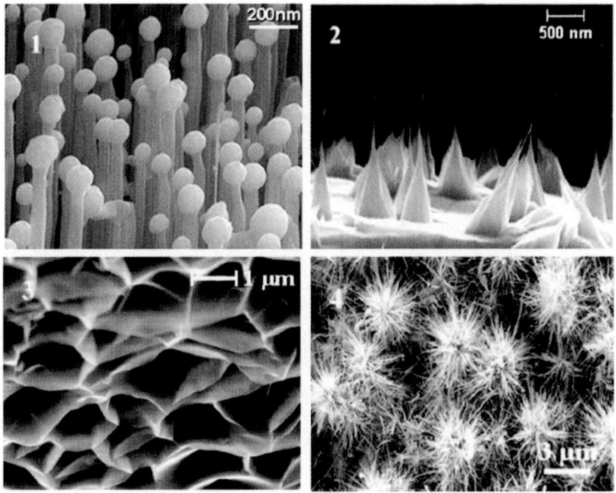

Fig. 13.15 SEM images of AlN samples for FE measurements: (1) nanorod array, (2) nanotip array, (3) nanoplatelets and (4) Si-doped nanoneedle array

tive electrometer (Keithley 6514) with an accuracy of 10^{-11} A. Figure 13.16a shows the measured FE current as a function of the applied voltage and the corresponding Fowler-Nordheim (F-N) plot (the inset) of 1D AlN nanotip array. It can be seen that the emission starts at 320 V, and the anode current reaches 81 nA as the voltage is increased to 470 V. Here, we define the turn-on field (E_{to}) and the threshold field (E_{thr}) as the electric fields required to produce a current density of $10\,\mu A/cm^2$ and $10\,mA/cm^2$, respectively.

The FE current-voltage characteristics were analyzed by using the Fowler- Nordheim (FN) equation (Taniyasu et al., 2004):

$$I = (1.54 \times 10^{-10} \beta^2 V^2 A/d^2 \phi) \exp(-6.83 \times 10^3 \phi^{3/2} d/\beta V)$$

where I is current, V is applied voltage, ϕ is work function of the emitting material, β is field enhancement factor, A and d are the area of emission and the distance between the anode and the cathode, respectively. The inset in Fig. 13.16a shows the FN plot of $\ln(I/V^2)$ vs I/V. The plot has an approximately linear relation within the measurement range, which confirms that the current indeed results from field emission. The field enhancement factor β can be calculated from the slope of the FN plot if the work function of the emitter is known.

Figure 13.16b shows the result of FE current versus time for a period over 240 min at a pressure of 5.0×10^{-9} torr under the fixed applied voltage of 1060 V for an initial current 77.89 μA. The average current and the standard deviation were calculated to be 77.53 and 0.57 μA, respectively. Since the emission area was 0.78 mm^2, the E_{to} and the E_{thr} for the AlN nanotip array were 4.7 V/μm and 10.6 V/μm, respectively. Taniyasu et al. (2004) obtained a field of 23 V/μm at 9.5 $\mu A/cm^2$ for Si-doped AlN film, which is about five times higher than our result. The E_{to} of AlN nanotip array is comparable to the reported values of carbon

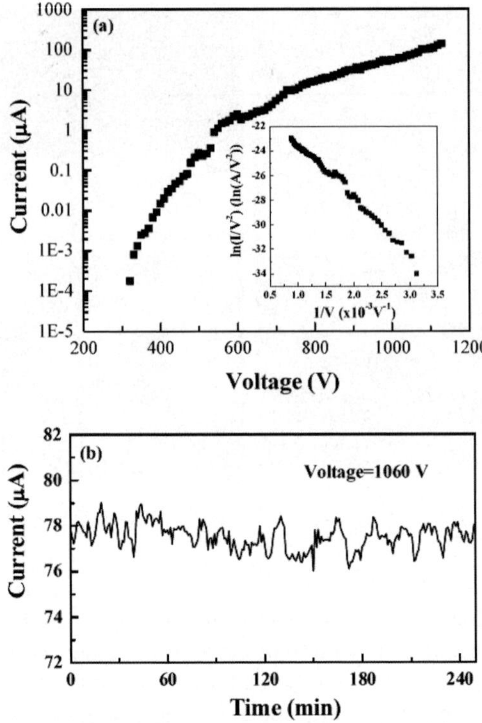

Fig. 13.16 The typical FE curves of 1D AlN array: (a) FE current as a function of the applied voltage, inset is the corresponding Fowler-Nordheim plot; (b) field emission current stability

nanotubes (Bonard et al., 1998; Jeong et al., 2001). The reason for the low turn-on electric field and high FE current density are probably due to single-crystalline structure and nanosize of the emitter. Many researchers ever reported that in order to avoid a field screening effect, aligned nanostructures with low density was favored (Jeong et al., 2001). The density of our AlN nanotip array is relatively low, approximately 10^6–10^7 tips/cm^2. Therefore, low value of E_{to} and high value of FE current are obtained.

The average current and the standard deviation were calculated to be 77.53 and 0.57 μA, respectively, for 4 hours FE stability test. The ratio of them is as low as 0.74% (Table 13.1), which proves the high current stability of Eiffel-tower-like AlN

Table 13.1 The field emission properties of four 1D AlN arrays obtained

Samples	1D arrays	E_{to} (V/μm)	E_{thr} (V/μm)	B	Fluctuation
1	Mushroom rod	8.8	19.2	565	<2%/1hour
2	Eiffel tip	4.7	10.6	1175–1889	0.74%/4hour
3	Honeycomb	3.2–5.0	7.8–12.1	1015–1785	2.7%/5hour
4	Si-doped needle	1.8	4.6	3271	<5%/5hour
Bonard et al. 1998	AlN film	34	84	——	3%/2hour
Taniyasu et al. 2004	AlN film	11	23	——	5.5%/1hour
Kasu et al. 2001	Carbon nantube	1.5–4.5	3.9–7.8	3600	>15%

nanotip. To the best of our knowledge, such a low fluctuation has not ever been reported for AlN material. Yin et al. (2005) reported a fluctuation of 7% for hierarchical AlN comb-like nanostructures. A fluctuation of 5.5% for heavily Si-doped AlN has been reported by Taniyasu et al. (2004). When an electric field is applied to an emitter, heat is generated at its sharp end because of high current density (Tondare et al., 2002). Therefore, the FE device will exhibit poor stability if the emitter has a poor thermal stability. The submicron-sized bases of our nanostructures have so large connection area that they can quickly transfer the heat from the tip to the Si substrate, so the tip can be effectively protected from being destroyed due to the superheat. Furthermore, the observed high stability of FE current is also attributed to the little change of the emission site, due to the strong bonding between the bases of AlN nanotips and the Si substrate.

The E_{to} value of the Si-doped AlN nanoneedle array in this work is lower than that previously reported (Table 13.1). The low E_{to}, high FE current, and high β of the Si-doped AlN nanoneedle array are mainly attributed to their sharp tips, high aspect ratio, and especially Si doping. The high aspect ratio and small radii of curvature of the AlN nanoneedles can generate a high local electric field at their tips, which decreases the FE potential barrier and so increases the FE current. On the other hand, according to Taniyasu et al. (2004), the Si dopant will form an impurity band just below the conduction band of AlN, therefore, the electrons for FE can easily tunnel through this shallow donor level to vacuum under the electric field. In addition, the Si doping remarkably lowers the resistance of AlN and increases the carrier concentration, which reduces the voltage drop along the nanoneedles and increases the effective field at their tips.

13.3.2.2 Photoluminescence (PL) of AlN Nanobelt Array

Figure 13.17 depicts the PL spectrum from the AlN nanobelt array (Tang et al., 2007a). It consists of a weak violet emission centered at $\sim 410\,\text{nm}$ (3.02 eV) and

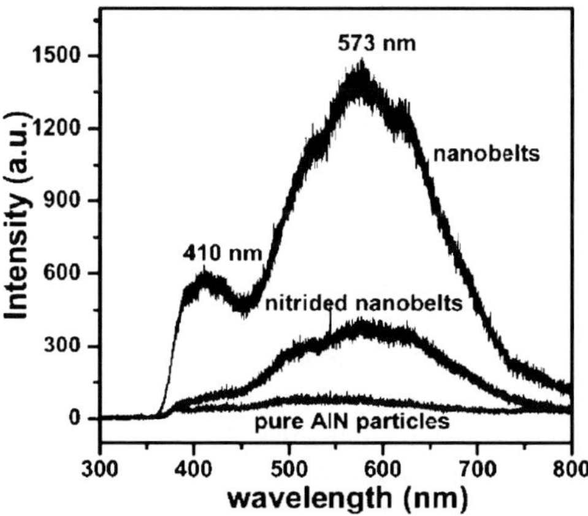

Fig. 13.17 Room-temperature PL spectra of the as-prepared AlN nanobelts, AlN nanobelts after NH$_3$ treatment and pure AlN particles

a strong green emission centered at $\sim 573\,nm$ (2.16 eV) with two shoulders of 523 and 619 nm. Apart from a slight difference of intensity, the main features of the PL spectrum are similar when the laser beam is focused on different regions of as-synthesized sample. Like the PL spectra from AlN nanowires and AlN nanocones reported by Zhao et al. (2005) and Liu et al. (2005), respectively, the direct band gap emission of the AlN nanobelt array can not be seen due to detection limitation. The observed weak violet emission of this nanostructure is similar to that of nanocrystalline AlN (420 nm) (Cao et al., 2000), in which such emission is related to the transition from shallow donor level to the excited state of the deep level. Lan et al. (1999) also reported emission at 405 nm (3.06 eV) from AlN nanopowders, and they attributed it to deep-level nitrogen vacancies and surface-luminescence.

The green luminescence band (2.16 eV) in this study is somewhat similar to the broad emission spectrum with peaks at 2.9, 2.3, and 2.2 eV from epitaxial AlN reported by Yim et al. (1973), and these peaks involve recombination at deep levels of unknown nature. The origin of this green emission in our AlN nanobelts is still not fully clear, but we believe that it is mainly originated from oxygen-related and structural defects associated with the peculiar nanostructures such as AlN nanowires and nanocrystalline AlN layer (Zhao et al., 2005; Siwiec et al., 1998). In this case, the nanobelts with small size and high specific surface area should also be easily doped with oxygen, which was detected by the TEM and EELS. The oxygen impurities will produce defects including oxygen point defects (O_N^+), nitrogen vacancies (V_N), and $V_{Al}^{3-} - 3 \times O_N^+$ complexes in this nanobelt (Mattila and Nieminen, 1996; Youngman and Harris, 1990). Thus the green emission may be resulted from the radioactive recombination of a photo-generated hole with an electron occupying the nitrogen vacancies and/or from the transition between the shallow level of V_N and the deep level of $V_{Al}^{3-} - 3 \times O_N^+$ complexes (Cao et al., 2000; Mattila and Nieminen, 1996). In addition, structural defects such as stacking faults and lattice dislocations may exist in our nanobelts according to the ripple-like contrast on the surface of some nanobelts in the TEM images because of strain resulting from the bending, which possibly induce the green emission because wurtzite AlN similar to wurtzite ZnO is a polar material.

Violet and green light emissions can be obtained from this AlN nanostructure, suggesting its potential applications in light emission display devices. And this nanostructure can be expected to serve as light sensors and field-emission devices, etc. which requires not only high purity but also good alignment.

13.3.2.3 Mechanical Properties and Thermal Expansion of an AlN Nanowires/Al Composite

The AlN-NWs were synthesized on quartz substrates by a vapor deposition method (Tang et al., 2007b). Figure 13.18a presents the typical scanning electron microscopy (SEM) image of the nanowires. The quartz substrate was covered by a large quantity of high-purity nanowires, and each of them has uniform diameter over its entire length. The nanowires have diameters in the range of 10–50 nm and lengths over several micrometers, and their aspect ratios are generally over 100. The

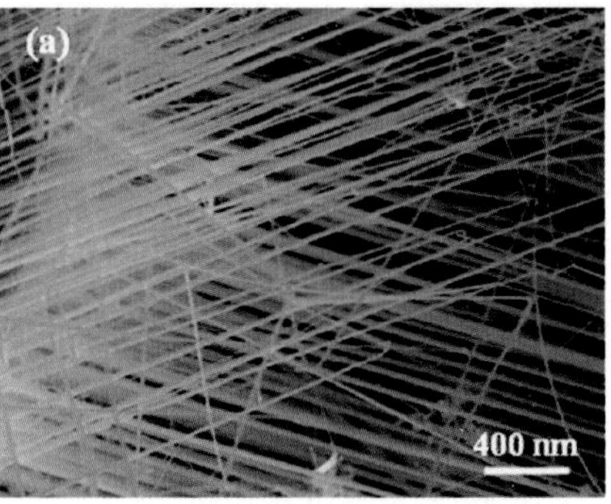

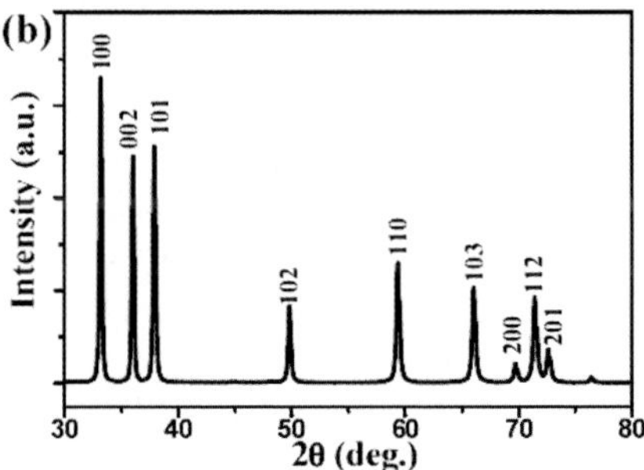

Fig. 13.18 (a) SEM image and (b) XRD pattern of the synthesized AlN nanowires

X-ray diffraction (XRD), shown in Fig. 13.18b, reveals the nanowires to exhibit the AlN wurtzite structure with the lattice constants of a = 3.114 Å and c = 4.979 Å (JCPDS: No. 25–1133).

The Al nanoparticles were produced by an active H_2 plasma evaporation technique. Figure 13.19 shows the schematic diagram of fabrication process of the AlN-NWs/Al composite. The three AlN-NWs/Al composites with AlN-NWs loading of 5, 10 and 15 vol.-% were prepared by ultrasonic mixing, and each composite powder was hot pressed in pure argon atmosphere under 50 MPa pressure at 700 °C to form a disk. For comparison, the Al matrix with the same size was also fabricated under the same conditions utilizing Al nanoparticles. The relative densities of all the AlN-NWs/Al composites and non-reinforced Al matrix were measured and calculated to be higher than 99%.

Figure 13.20 shows the representative SEM image of the polished AlN-NWs/Al composite containing 15 vol.-% AlN-NWs after etched by 1% hydrofluoric acid. AlN-NWs can be seen to be embedded in the continuous Al matrix exposing both their cross-sectional and the longitudinal views, indicating a non-directional alignment of the nanowires in the compacted specimen. There is no agglomeration of AlN nanowires in composite specimens due to the uniform dispersion of AlN-NWs in Al nanoparticles during mixing because of the minute difference of density between the AlN ($\sim$3.2 g/cm^3) and Al ($\sim$2.7 g/cm^3). The composite material appears to be macroscopically homogeneous and free of porosity owing to the high

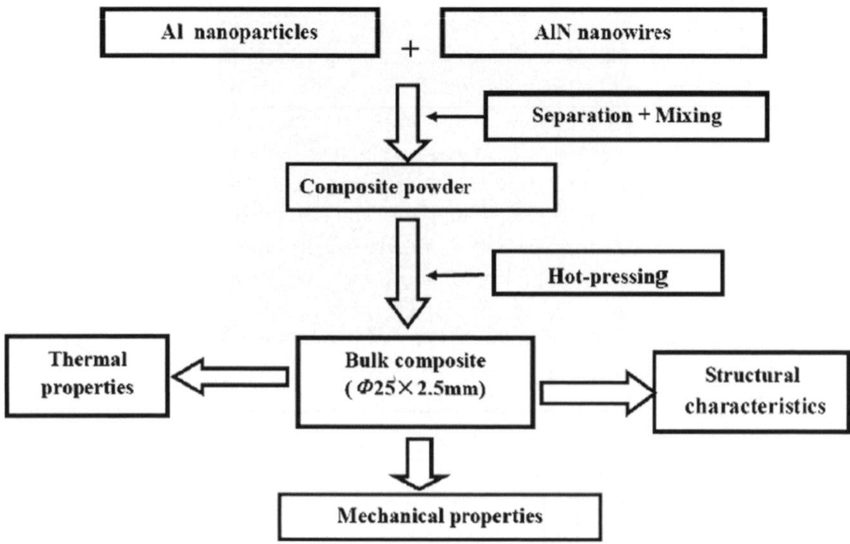

Fig. 13.19 The schematic diagram of fabrication process of the AlN-NWs/Al composite

pressure employed during fabrication to make the molten Al to fill in the mould from edge to center completely.

It is known that interfacial characteristics are of great importance to the performance and properties of composites (Dhingra and Fishman, 1986). Having a good interface, composites can effectively transfer the external load from the matrix to the reinforcement. And the interface mainly depends on the wetting between the matrix and reinforcement. The adhesion between the reinforcement and the matrix must be adequate, so that the composite achieves good shear and off-axis properties

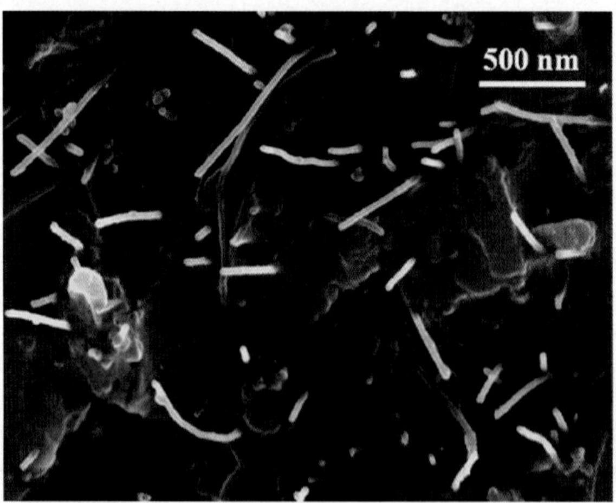

Fig. 13.20 The representative SEM image of the polished 15 vol.-% AlN-NWs composite after etched by 1% hydrofluoric acid

normally required by reinforced metals (Piggott, 1980). On the contrary, microvoids will form on the interface if the wetting is inadequate, which will result in the decrease of interfacial bonding strength. Additionally, harmful chemical reactions between the reinforcement and the matrix must be limited to prevent the reinforcement from being weakened (Dhingra and Fishman, 1986). So, it is necessary to do further investigation on the interface of AlN-NWs/Al composites.

Figure 13.21a and b are the TEM images of the composite containing 15 vol.-% AlN-NWs. AlN-NWs can be clearly observed in the matrix, indicating that the nanowires are not broken by the high temperature and pressure consolidation. An enlarged HRTEM image of the interfacial region is shown in Fig. 13.21c. It reveals that the interface between AlN nanowire and Al matrix is clean and bonded well. No porosity or interfacial reaction product is observed at the nanowire/matrix boundary by the HRTEM observations. We assume that the improvement of mechanical properties should be attributed to the strongly bonded interfaces, which can efficiently transfer stress from the matrix to the reinforcement.

Mechanical properties including tensile strength, Young's modulus and hardness of the composites are improved with AlN-NWs volume fraction changing from 5 to 15 vol.-%, and the maximum tensile strength, Young's modulus and hardness of the composite are about 5, 3 and 2 times higher than those of Al matrix. Meanwhile, AlN-NWs effectively decrease the coefficient of thermal expansion (CTE) of Al matrix, and the CTE of the AlN-NWs/Al composite with 15 vol.-% AlN-NWs is about one half of that of the Al matrix. The results obtained suggest that the

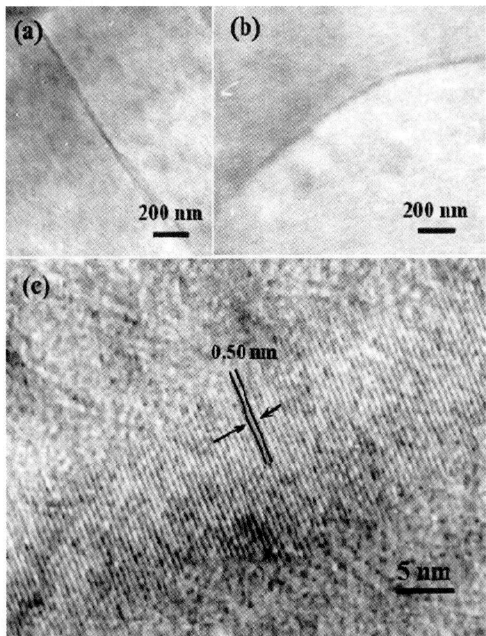

Fig. 13.21 (a), (b) TEM images of the composite containing 15 vol.-% AlN-NWs, and (c) an enlarged HRTEM image of the interfacial region

AlN nanowire is an ideal reinforcement for optimizing the mechanical and thermal properties of metal matrix composites.

13.4 Conclusion and Outlook

Various quasi-one-dimensional (1D) nitride nanostructures including BN nanotubes and nanofibers, AlN nanorods, AlN nanotips, AlN nanobelts, and AlN nanofibers are synthesized. Especially, the single-crystalline AlN mushroom-like nanorod array, Eiffel-tower-shape nanotip array, honeycomb-like network of vertically aligned nanoplatelets and flowerlike Si-doped AlN nanoneedle array are selectively synthesized by an improved chemical vapor transport and condensation method and a mobile nitrogen arc discharge method, respectively. Field emission (FE) measurements for AlN aligned 1D nanostructures show that they have very low turn-on voltages, high emission currents and small FE current fluctuations. It is revealed the morphology of the nanostructures can be tailored to optimize their FE properties and make them promising field emitters. Moreover, a strong green light emission is obtained from AlN nanobelts. Mechanical properties and thermal expansion of a composite of AlN nanowires and nanocrystalline Al implies that the AlN nanowire is also an ideal reinforcement for improving the mechanical and thermal properties of metal matrix composites.

Acknowledgments The authors thank Drs. F. Li, Z.G. Chen, B. Wu, and J.F. Wu, Mr. B.J. Xia and K.Y. Hu for their kind help. The project was supported by National Natural Science Foundation of China (Grant No. 50371083)

References

Benjamin, M. C., Wang, C., Davis, R. F., and Nemanich, R. J. 1994, Observation of a negative electron-affinity for heteroepitaxial AlN on alpha(6H)-SiC(0001), *Applied Physics Letters*, 64, pp. 3288–3290.

Bonard, J. M., Salvetat, J. P., Stockli, T., de Heer, W. A., Forro, L. and Chatelain, A. 1998, Field emission from single-wall carbon nanotube films. *Applied Physics Letters*, 73, pp. 918–920.

Bradshaw, S. M. and Spicer, J. L. 1999, Combustion synthesis of aluminum nitride particles and whiskers. *Journal of the American Chemical Society*, 82, pp. 2293–2300.

Cao, Y. G., Chen, X. L., Lan, Y. C., Li, J. Y., Xu, Y. P., Xu, T., Liu, Q. L. and Liang, J. K. 2000, Blue emission and Raman scattering spectrum from AlN nanocrystalline powders. *Journal of Crystal Growth*, 213, pp. 198–202.

Chedru, M., Chermant, J. L. and Vicens, J. 2001, Thermal properties and Young's modulus of Al-AlN composites. *Journal of Materials Science Letters*, 20, pp. 893–895.

Chen, G. D., Smith, M., Lin, J. Y., Jiang, H. X., Wei, S. H., Khan, M. A. and Sun, C. J. 1996, Fundamental optical transitions in GaN, *Applied Physics Letters*, 68, pp. 2784–2786.

Chen, X. and Gonsalves, K. E. 1997, Synthesis and properties of an aluminum nitride/polyimide nanocomposite prepared by a nonaqueous suspension process. *Journal of Materials Research*, 12, pp. 1274–1286.

Dhingra, A. K. and Fishman, S. G. 1986, *Interfaces in Metal-Matrix Composites* (New Orleans, Metallurgical Society Inc.), p. 211.

Geiger A. L. and Jackson, M. 1989, Low-expansion MMCs boost avionics. *Advanced Materials & Processes*, 136, pp. 23–30.

Grabowski, S. P., Schneider, M., Nienhaus, H., Monch, W., Dimitrov, R., Ambacher, O. and Stutzmann, M. 2001, Electron affinity of AlxGa1-xN(0001) surfaces. *Applied Physics Letters*, 78, pp. 2503–2505.

Haber, J. A., Gibbons, P. C. and Buhro, W. E. 1997, Morphological control of nanocrystalline aluminum nitride: Aluminum chloride-assisted nanowhisker growth, *Journal of the American Chemical Society*,119, pp. 5455–5456.

Haber, J. A., Gibbons, P. C. and Buhro, W. E. 1998, Morphologically selective synthesis of nanocrystalline aluminum nitirde. *Chemistry of Materials*, 10, pp. 4062–4071.

He, J. H., Yang, R., Chue, Y. L., Chou, L. J., Chen, L. J. and Wang, Z. L. 2006, Aligned AlN nanorods with multi-tipped surfaces - Growth, field-emission, and cathodoluminescence properties. *Advanced Materials*, 18, pp. 650–654.

Huang, J. L. and Li, C. H. 1994, Microstructure and mechanical properties of aluminum nitride aluminum composite. *Journal of Materials Research*, 9, pp. 3153–3159.

Iijima, S. 1991, Helical microtubules of graphitic carbon. *Nature*, 354, pp. 56–58.

Inoue, A., Nosaki, K., Kim, B. G., Yamaguchi, T. and Masumoto, T. 1993, Mechanical strength of ultra-fine Al-AlN composites produced by a combined method of plasma-alloy reaction, spray deposition and hot pressing. *Journal of Materials Science*, 28, pp. 4398–4404.

Jeong, J. S., Lee, J. Y., Lee, C. J., An, S. J. and Yi, G. C. 2004, Synthesis and characterization of high-quality In_2O_3 nanobelts via catalyst-free growth using a simple physical vapor deposition at low temperature. *Chemical Physics Letters*, 384, pp. 246–250.

Jeong, S. H., Hwang, H. Y., Lee, K. H. and Jeong, Y. 2001, Template-based carbon nanotubes and their application to a field emitter. *Applied Physics Letters*, 78, pp. 2052–2054.

Kasu, M. and Kobayashi, N. 2001, Spontaneous ridge-structure formation and large field emission of heavily Si-doped AlN. *Applied Physics Letters*, 78, pp.1835–1837.

Lai S. W. and Chung, D. D. 1994, Superior high-temperature resistance of aluminum nitride particle-reinforced aluminum compared to silicon-carbide or alumina particle-reinforced aluminum. *Journal of Materials Science*, 29, pp. 6181–6198.

Lan, Y. C., Chen, X. L., Cao, Y. G, Xu, Y. P., Xun, L. D., Xu, T. and Liang, J. K. 1999, Low-temperature synthesis and photoluminescence of AlN. *Journal of Crystal Growth*, 207, pp. 247–250.

Li, Y. B., Bando, Y. and Golberg, D. 2003, MoS_2 nanoflowers and their field-emission properties. *Applied Physics Letters*, 82, pp. 1962–1964.

Liang, C., Shimizu, Y., Sasaki, T., Umehara, H. and Koshizaki, N. J. 2004, Au-mediated growth of wurtzite ZnS nanobelts, nanosheets, and nanorods via thermal evaporation. *Journal of Physics Chemical B*, 108, pp. 9728–9733.

Liang, Y. J. and Che M. X. 1993, *Handbook for Thermodynamic Data of Inorganic Compounds* (China, Press of Northeast University).

Liu, B. D., Bando, Y., Tang, C. C., Golberg, D., Xie, R. G. and Sekiguchi T. 2005a, Synthesis and optical study of crystalline GaP nanoflowers, *Applied Physics Letters*, 86, p. 083107.

Liu, C., Hu, Z., Wu, Q., Wang, X. Z., Chen, Y., Sang, H., Zhu, J. M., Deng, S. Z. and Xu, N. S. 2005b, Vapor-solid growth and characterization of aluminum nitride nanocones, *Journal of the American Chemical Society*, 127, pp. 1318–1322.

Liu, J., Zhang, X., Zhang, Y., He, R. and Zhu J. 2001, Novel synthesis of AlN nanowires with controlled diameters. *Journal of Materials Research*, 16, pp. 3133–3138.

Mattila, T. and Nieminen, R. M. 1996, *Ab initio* study of oxygen point defects in GaAs, GaN, and AlN. *Physical Review B*, 54, pp. 16676–16682.

Morales, A. M. and Lieber, C. M. 1998, A laser ablation method for the synthesis of crystalline semiconductor nanowires. *Science*, 279, pp. 208–211.

Nakamura, S. 1998, The roles of structural imperfections in InGaN-Based blue light-emitting diodes and laser diodes. *Science*, 281, pp. 956–961.

Pan, Z. W., Dai, Z. R. and Wang, Z. L. 2001, Nanobelts of semiconducting oxides. *Science*, 291, pp. 1947–1949.

Piggott, M. R. 1980, *Load-Bearing Fibre Composites* (New York: Pergamon Press Inc.), Ch. 10, p. 286.

Ronning, C., Banks, A. D., McCarson, B. L., Schlesser, R., Sitar, Z., Davis, R. F., Ward, B. L. and Nemanich, R. J. 1998, Structural and electronic properties of boron nitride thin films containing silicon. *Journal of Applied Physics*, 84, pp. 5046–5051.

Shi, S. C., Chen, C. F., Chattopadhyay, S., Lan, Z. H., Chen, K. H. and Chen, L. C. 2005, Growth of single-crystalline wurtzite aluminum nitride nanotips with a self-selective apex angle. *Advanced Functional Materials*, 15, pp. 781–786.

Siwiec, J., Sokolowska, A., Olszyna, A., Dwilinski, R., Kaminska, M. and Hrabowska, J. K. 1998, Photoluminescence properties of nanocrystalline, wide band gap nitrides (C_3N_4, BN, AlN, GaN), *Nanostructural Materials*, 10, pp. 625–634.

Sugino, T., Hori, T., Kimura, C. and Yamamoto, T. 2001, Field emission from GaN surfaces roughened by hydrogen plasma treatment. *Applied Physics Letters*, 78, pp. 3229–3231.

Suzuki, M., Uenoyama, T. and Yanase, A. 1995, First-principles calculations of effective- mass parameters of AlN and GaN, *Physical Review B*, 52, pp. 8132–8139.

Tang, Y. B., Cong, H. T., Chen, Z. G. and Cheng, H. M. 2005a, An array of Eiffel-tower-shape AlN nanotips and its field emission properties. *Applied Physics Letters*, 86, p. 233104.

Tang, Y. B., Cong, H. T., Wang, Z. M. and Cheng H. M. 2005b, Synthesis of rectangular cross-section nanofibers AlN by chemical vapor deposition. *Chemical Physics Letters*, 416, pp. 171–175.

Tang, Y. B., Cong, H. T., Zhao, Z. G. and Cheng, H. M., 2005c, Field emission from AlN nanorod array. *Applied Physics Letters*, 86, p. 153104.

Tang, D. M., Liu, C., Cong, H. T. and Cheng, H. M. 2006a, Platelet boron nitride nanowires. *Nano*, 1, pp. 65–71.

Tang, Y. B., Cong, H. T. and Cheng, H. M. 2006b, Field emission from honeycomb-like network of vertically aligned AlN nanoplatelets, *Applied Physics Letters*, 89, p. 093113.

Tang, Y. B., Cong, H. T., Wang, Z. M. and Cheng, H. M. 2006c, Catalyst-seeded synthesis and field emission properties of flowerlike Si-doped AlN nanoneedle array. *Applied Physics Letters*, 89, pp. 253112.

Tang, Y. B., Cong, H. T., Li, F. and Cheng, H. M. 2007a, Synthesis and photoluminescent property of AlN nanobelt array on Si substrate. *Diamond and Related Materials*, 16, pp. 537–541.

Tang, Y. B, Liu, Y., Sun, C. H. and Cong, H. T. 2007b, AlN nanowires for Al-based composites with high strength and low thermal expansion. *Journal of Materials Research* (accepted).

Taniyasu, Y., Kasu, M. and Makimoto, T. 2004, Field emission properties of heavily Si-doped AlN in triode-type display structure. *Applied Physics Letters*, 84, pp. 2115–2117.

Tanniyasu, Y., Kasu, M. and Makimoto, T. 2006, An aluminium nitride light-emitting diode with a wavelength of 210 nanometres. *Nature*, 441, pp. 325–328.

Tondare, V. N., Balasubramanian, C., Shende, S. V., Joag, D. S., Godbole, V. P., Bhoraskar, S. V. and Bhadbhade, M. 2002, Field emission from open ended aluminum nitride nanotubes. *Applied Physics Letters*, 80, pp. 4813–4815.

Vicens, J., Chedru, M., Cubero, H. and Chermant, J. L. 2002, Effects of AlN additions and heat treatments on the compression behavior of Al-AlN composites. *Journal of Materials Science Letters*, 21, pp. 1505–1508.

Vurgaftman, I., Meyer, J. R. and Ram-Mohan, L. R. 2001, Band parameters for III-V compound semiconductors and their alloys. *Journal of Applied Physics*, 89, pp. 5815–5875.

Ward, B. L., Nam, O. H., Hartman, J. D., English, S. L., McCarson, B. L., Schlesser, R., Sitar, Z., Davis, R. F. and Nemanish, R. J. 1998, Electron emission characteristics of GaN pyramid arrays grown via organometallic vapor phase epitaxy. *Journal of Applied Physics*, 84, pp. 5238–5242.

Wei, S. H. and Zunger, A. 1996, Valence band splittings and band offsets of AlN, GaN, and InN. *Applied Physics Letters*, 69, pp. 2719–2721.

Wu, Q., Hu, Z., Wang, X., Lu, Y., Chen, X., Xu, H. and Chen Y. 2003a, Synthesis and characterization of faceted hexagonal aluminum nitride nanotubes. *Journal of the American Chemical Society*, 125, pp. 10176–10177.

Wu, Q., Hu, Z., Wang, X. Z. and Chen, Y. 2003b, Synthesis and optical characterization of aluminum nitride. *Journal of Physical Chemistry B*, 107, pp. 9726–9729.

Wu, Q., Hu, Z., Wang, X., Hu, Y., Tian, Y. and Chen, Y. 2004, A simple route to aligned AlN nanowires. *Diamond and Related Materials*, 13, pp. 38–41.

Xia, Y., Yang, P., Sun, Y., Wu, Y., Mayers, B., Gates, B., Yin, Y., Kim, F. and Yan, H. 2003, One-dimensional nanostructures: synthesis, characterization, and applications. *Advanced Materials*, 15, pp. 353–389.

Xu, C. X., Sun, X. W. and Chen, B. J. 2004, Field emission from gallium-doped zinc oxide nanofiber array. *Applied Physics Letters*, 84, pp. 1540–1542.

Yim, W. M., Stofko, E. J., Zanzucchi, P. J., Pankove, J. L., Ettenberg, M. and Gilbert, S. L. 1973, Epitaxially grown AlN and its optical band gap. *Journal of Applied Physics*, 44, pp. 292–296.

Yin, L. W., Bando, Y., Zhu, Y. C., Li, M. S., Li, Y. B. and Golberg, D. 2005a, Growth and field emission of hierarchical single-crystalline wurtzite AlN nanoarchitectures. *Advanced Materials*, 17, pp. 110–114.

Yin, L. W., Bando, Y., Zhu, Y. C., Li, M. S., Tang, C. C. and Golberg, D. 2005b, Single-crystalline AlN nanotubes with carbon-layer coatings on the outer and inner surfaces via a multiwalled-carbon-nanotube-template-induced route. *Advanced Materials*, 17, pp. 213–217.

Youngman, R. A. and Harris, J. H. 1990, Luminescence studies of oxygen-related defects in aluminum nitride. *Journal of American Ceramic Society*, 73, pp. 3238–3246.

Zhang, J. and Zhang, L. 2002, Intensive green light emission from MgO nanobelts, *Chemical Physics Letters*, 363, pp. 293–297.

Zhang, Q., Chen, G., Wu, G., Xiu, Z. and Luan, B. 2003, Property characteristics of a AlNp/Al composite fabricated by squeeze casting technology. *Materials Letters*, 57, pp. 1453–1458.

Zhang, Q., Wu, G., Sun D. and Luan, B. 2002, Study on the thermal expansion and thermal cycling of AlNp/Al composites. *Journal of Materials Science & Technology*, 57, pp. 63–65.

Zhang, Y. J., Liu, J., He, R. R., Zhang, Q., Zhang, X. Z. and Zhu, J. 2001, Synthesis of aluminum nitride nanowires from carbon nanotubes. *Chemistry of Materials*, 13, pp. 3899–3905.

Zhao, Q., Zhang, H., Xu, X., Wang, Z., Xu, J., Yu, D., Li, G. and Su, F. 2005, Optical properties of highly ordered AlN nanowire arrays grown on sapphire substrate, *Applied Physics Letters*, 86, p. 193101.

Chapter 14
Electron Energy-Loss Spectroscopy for Nannomaterials

C. H. Chen* and M. W. Chu

Electron energy-loss spectroscopy (EELS) using high-energy incident electrons has been known to be a powerful tool for studies of electronic structure in solids. The main energy-loss mechanisms in thin solid samples are due to single-particle excitations such as interband and core-level transitions, and collective excitations such as bulk plasmons and surface plasmons. In EELS, the intensity of inelastically scattered incident electrons could be measured as a function of energy-losses and scattering angles (reciprocal space) under the condition of parallel beam illumination. Energy-loss spectra obtained as a function of scattering angles can yield the dispersion relations, i.e., the excitation energy vs. wave vector, of the corresponding excitations. The angular (wave vector) dependence of the EELS spectra plays an important role for us to fully understand the elementary excitations in solids.

High resolution spatially-resolved EELS spectra can be obtained from different areas of interest in a sample when a convergent beam electron nanoprobe is utilized. In this case, the spatial resolution is determined primarily by the electron probe size. This is readily accomplished when EELS is performed in conjunction with a modern scanning transmission electron microscope (STEM) which can deliver an electron probe ~ 0.2 nm or smaller. The combination of EELS with a STEM is a very powerful tool for the studies of local electronic structures on an atomic-scale for complex materials. This is particularly suitable for studies of nanomaterials in which inhomogeneity of sizes, shapes and structures often can not be avoided. The combination of STEM/EELS can circumvent this difficulty and allows us to study each nano-object individually. It should be noted, however, that the use of nano-probes under convergent beam conditions for spatially resolved EELS inevitably collected electrons inelastically scattered through a wide angle. The measured scattering cross-section is, therefore, angle-integrated in this case.

In most solids, the character of collective plasmon excitations normally can be easily distinguished from single particle excitations such as interband and core-level transitions. This is particularly true for metals well described by the nearly free electron-gas model in which the energies of these two types of excitations are well separated. Aluminum which exhibits a bulk plasmon peak at 15 eV and an interband transition at 1.5 eV is an excellent example in this category. The situation,

---

* Center for Condensed Matter Sciences, National Taiwan University, Taipei, Taiwan

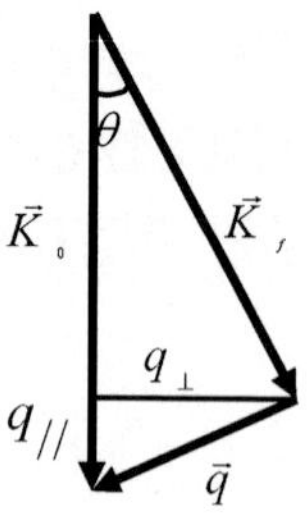

Fig. 14.1 Scattering kinematics for EELS

however, becomes more difficult for transition and nobles metals where clear distinction between collective and single particle effects can not be made unambiguously. This is mainly because the oscillator strengths for single particle transitions involving d electrons are distributed over a rather wide energy range. The intimate mixture of collective and single particle excitations presents new challenges to the interpretation of the EELS spectra in this class of metals, especially for studies of nanomaterials. We'll first review some basic physics involved in EELS measurements and this is followed by exemplifications of EELS studies of thin aluminum films and gold nanoparticles.

A simple scattering kinematics is shown in Fig. 14.1. A parallel beam of incident high energy electron scattered by an inelastic event resulting in a loss of energy $\hbar\omega$ will also undergo a corresponding momentum transfer q. For typical incident electron energies in a TEM ($E_0 = 200\,\mathrm{keV}$) and energy transfers (typically less than $2\,\mathrm{keV}$), the momentum transfer q can be approximated by,

$$q = K_0\sqrt{\theta^2 + \theta_E^2}, \text{ where } \theta_E = \hbar\omega/2E_0$$

The momentum transfer q is essentially perpendicular to incident beam direction under normal scattering conditions for bulk excitations since $\theta \in \theta_E$. The differential inelastic scatterings cross-section of a material characterized by a complex dielectric function $\varepsilon\,(q, \omega)$ is given by (Pines, 1963; Raether, 1965),

$$\frac{d^2\sigma}{d\omega d\Omega} \approx \frac{1}{q^2}\mathrm{Im}\left[\frac{-1}{\varepsilon\,(q, \omega)}\right], \text{ where } \varepsilon = \varepsilon_1 + i\varepsilon_2$$

and

$$\mathrm{Im}\left[\frac{-1}{\varepsilon\,(q, \omega)}\right] = \frac{\varepsilon_2}{\varepsilon_1^2 + \varepsilon_2^2}$$

is the so-called energy-loss function. Well-defined maximum of the energy-loss function occurs at a particular frequency and wave-vector when the real part of the dielectric function is near zero and the corresponding imaginary part is small. In the long wavelength limit when $q \sim 0$, the dielectric function for a nearly free electron gas, such as aluminum, can be well described by the Drude model,

$$\varepsilon(\omega) = 1 - \frac{\omega_P^2}{\omega\,(\omega + i\gamma)}, \quad \text{where} \quad \omega_P = \sqrt{\frac{4\pi n e^2}{m}}$$

is the plasmon frequency and γ is the damping factor. The energy-loss function is then sharply peaked at the plasma frequency where the real part of the dielectric function crosses the frequency axis with a positive slope. The predominant inelastic scattering cross-section is therefore due to the excitations of bulk plasmons. At finite q, the random phase approximation for a nearly free electron gas yields a plasmon energy varying quadratically in q, $\omega = \omega_P + \alpha\left(\hbar/m\right)q^2$ where $\alpha = 3E_F/5\hbar\omega_P$ is the plasmon dispersion coefficient and E_F is the Fermi energy.

Let's now look at some experimental EELS results for aluminum. We show in Fig. 14.2 a couple of EELS spectra obtained at two different momentum transfers. At $q = 0$, we observe a strong plasmon resonance at 15 eV consistent with earlier experimental results and a secondary peak at 30 eV due to plural scatterings (Batson et al., 1976). Unlike inelastic x-ray scattering, multiple scattering is a common occurrence in EELS due to the much stronger interactions with solids unless extremely thin samples ($\sim$ a few nanometers in thickness) are used. Removal of multiple scattering contributions in the EELS spectra is often necessary in order to compare with theoretical calculations. In addition to the bulk plasmon excitations, it is noted that a surface plasmon peak located at 7.0 eV is also present. Surface plasmons are charge density fluctuations localized at the sample surfaces. The condition for the existence of surface plasmon is that ε_1 must be negative and the resonance occurs when $\varepsilon_1 = -1$. For a clean aluminum surface, the surface plasmon energy is then determined to take place at 10.6 eV. Nevertheless, the presence of a thin natural oxide layer on the surface is expected to push the surface plasmon energy down to 7.0 eV as observed (Batson et al., 1976). The dispersive nature of bulk plasmon is demonstrated clearly in the spectra taken with $q = 1.18$ A^{-1}. The plasmon peak has now moved upward to an energy $\sim$ 18.5 eV and becomes much broader as expected. The weaker and non-dispersive peak at 15 eV is the spurious peak due to multiple scatterings involving a plasmon and phonons. It is noted that the intensity of surface plasmon peak has decreased drastically as the scattering angle increases and becomes unobservable at this momentum transfer. For aluminum,

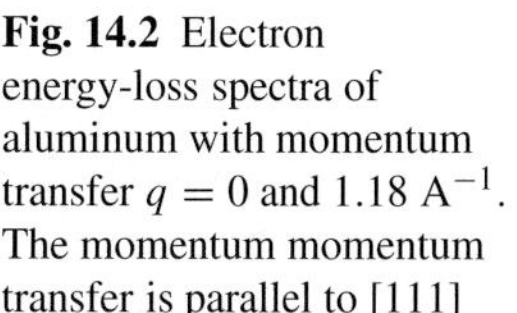

Fig. 14.2 Electron energy-loss spectra of aluminum with momentum transfer $q = 0$ and 1.18 A^{-1}. The momentum momentum transfer is parallel to [111]

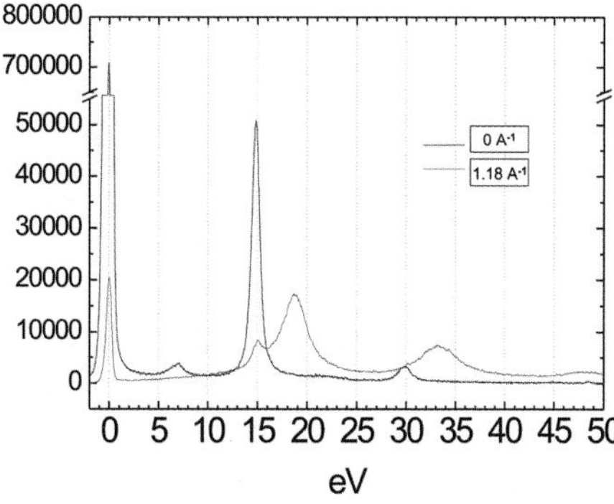

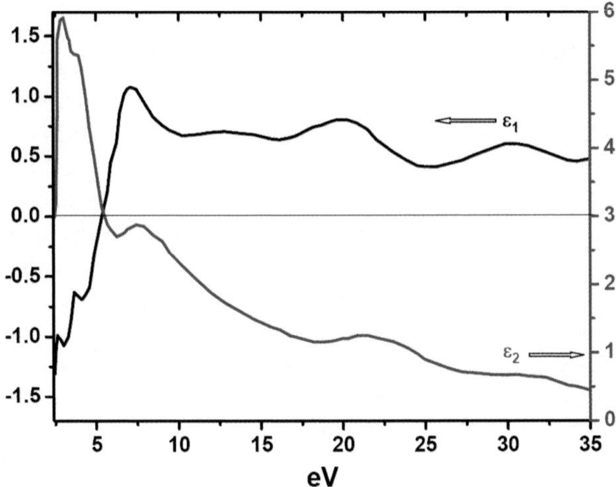

Fig. 14.3 Frequency dependent dielectric function of Au

the strongest interband transition occurs at 1.5 eV (Chen and Silcox, 1977), far below the plasmon excitation, and its effect on the collective plasmon peak is negligible.

The situation for gold metal is totally different. This difference is clearly demonstrated in the frequency dependent dielectric function of Au (Schlueter, 1972; Cooper and Ehrenreich, 1965) shown in Fig. 14.3. It certainly cannot be approximated by the simple Drude model discussed above for aluminum. From the dielectric function, bulk plasmon is expected to occur at 5.0 eV where the ε_1 becomes zero with a positive slope. However, this bulk plasmon resonance is strongly damped due to the high values of ε_2, which is caused by interband transitions in the propinquity. Below 5 eV, ε_1 is negative exhibiting the highly reflective metallic characteristics of Au and the presence of surface plasmon peak at 2.4 eV in the visible spectral regime is well known. There exist several broad maxima in ε_2 with energies higher than 10 eV signifying the widely distributed oscillator strengths. It is also noted that, in this deep-UV spectral regime, ε_1 has becomes positive with poorly defined broad resonances occurring in tandem with the broad maxima in ε_2. The strongest peaks in EELS of Au are actually observed in this spectral regime, consistent with the energy-loss function calculated from the dielectric function. It is in this deep-UV spectral regime that we observe the unexpected excitations of surface plasmon-like resonances in Au nanoparticles using EELS with a 0.2 nm electron probe in near-field geometry. The 0.2 nm electron probe is generated in a FEI field-emission STEM, Tecnai F20, operated at 200 kV.

Figure 14.4 shows a set of corresponding EELS spectra taken by placing the 0.2 nm electron probe in various locations inside or around the nanopartilces as indicated. Spectra #1 and 5 were obtained from the center of the nanoparticles and are predominated by the characteristic bulk excitations at 17, 25.5 and 33.5 eV. The free electron bulk plasmon energy for the s and s plus d electrons in Au is calculated to be ∼9 and 33 eV, respectively. We note that there is no any indication of the 9 eV

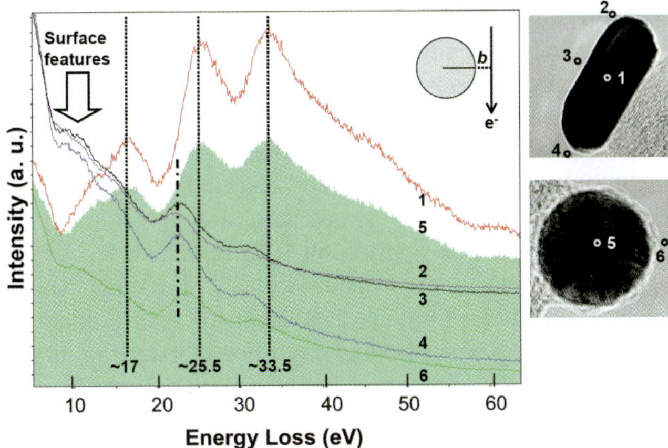

Fig. 14.4 Electron energy-loss spectra obtained from gold nanoparticles. Spectra #1 and 5 predominated by bulk excitations were taken from the center of the nanoparticles and spectra #2, 3, 4 and 6 reflecting the surface excitations were obtained by placing the 0.2 nm electron probe at a distance 1 nm away from the particle surface. The diameter of the rod-like and spherical nanoparticle is 10 and 13 nm, respectively

peak in spectra #1 and 5 that can be associated with the bulk plasmon excitation due to s electrons. The physical origin of the 17 eV peak is unclear, although interband transition could be responsible. The 33.5 eV peak may be related to the bulk plasmon excitations of s plus d electrons, but the significant oscillator strengths of interband transition are still present at this energy regime and the effective number of valence electrons participating the 33.5 eV excitation are still far below the total 11 electrons in the s and d bands.

Spectra #2, 3, 4, and 6 were taken with the probe placed outside the nanoparticles and $\sim$ 1 nm away from the surface. In this near-field geometry, only surface excitations are possible due to the evanescent wave-fields of excitations confined at the surface and no bulk excitations are expected. The most noticeable feature in these spectra is the appearance of the excessive intensities $\sim$ 10 eV, absent in the bulk spectra of #1 and 5. Furthermore, all the bulk peaks at 17, 25.5 and 33.5 eV shown in the spectra #1 and 5 appear to have shifted to lower energies by about 2 eV, but slight variations of the peak positions in the spectra due to different geometrical configurations around the probe positions are discernible. The appearance of these surface plasmon-like features above 10 eV is not expected in view of the positive dielectric function ε_1 shown in Fig. 14.3, which indicates that Au is no longer metallic and behaves like an insulator in this spectral regime. Conventional surface plasmons, as mentioned earlier, are known to exist only in the metallic spectral regime where ε_1 is negative. It should be noted that the surface plasmon-like excitations above 10 eV are closely linked to the presence of interband transition maxima in the imaginary part of the dielectric function shown in Fig. 14.3. A positive value of ε_1 alone is not sufficient to warrant the appearance of these surface excitations. For instance, although ε_1 becomes positive above the bulk plasmon energy at 15 eV in aluminum, no surface excitations similar to what we just described for Au nanoparticles are expected since there are no interband transitions of any significant oscillator strength in this spectral regime, which will link to resonance peaks in ε_2.

The intricate mixing of single particle characteristics in a collective Surface excitation calls for a new thinking of elementary excitations in solids. A new physical picture regarding the presence of surface plasmon-like excitations in the non-metallic spectral regime seems necessary to grasp the full significance of these occurrences observed in electron energy loss spectroscopy.

In this manuscript we show an application of EELS/STEM for Au nanoparticles. The ability to obtain local electronic structure with a 0.2 nm spatial resolution makes this technique ideal for studies for individual nanoparticle in a, otherwise, inhomogeneous nanomaterials. One can also explore different excitation modes governed by the local geometry of the nanoparticle by placing the electron nanoprobe at the particular points of interest. Moreover, coupled excitation modes between nanoparticles can also be studied. It is expected some coupled modes can only be excited by EELS and not by any optical method. EELS can also be used to measure bandgaps of semiconductors and insulators. Possible local bandgap fluctuations in nanomaterials, which would be an impossible task for most experimental measurements, can be readily explored by the EELS/STEM technique.

References

Batson, P. E., Chen, C. H. and Silcox, J., 1976, *Phys. Rev. Lett.* **37**, p. 937.
Chen, C. H. and Silcox, J., 1977, *Phys. Rev.* **B16**, p. 4246.
Cooper, B. R. and Ehrenreich, H., 1965, *Phys. Rev.* **138**, p. A494.
Pines, D., 1963, *Elementary Excitations in Solids* (New York-Amsterdam, W. A. Benjamin Inc.)
Raether, H., 1965, *Mod. Phys.* **38**, p. 85.
Schlueter, M., 1972, *Z. Physik* **250**, p. 87.

Chapter 15
Fabrication of Photovoltaic Devices by Layer-by-Layer Polyelectrolyte Deposition Method

Wai Kin Chan*, Ka Yan Man, Kai Wing Cheng and Chui Wan Tse

15.1 Introduction

Layer-by-layer (LbL) electrostatic self-assembly is a versatile method for preparation of polymer multilayer thin films (Decher and Hong, 1991). This method has been employed in the preparation of different functional polymer films for a variety of applications. In this paper, we report the fabrication of different photovoltaic devices using the layer-by-layer polyelectrolyte deposition method. Compared to other film forming techniques in polymers (e.g. spin-coating), this method can yield multilayer thin films with highly reproducible thickness (resolution up to few nm) easily. The control of device thickness is a very important issue in the fabrication of photovoltaic cells because film thickness will affect the device performance strongly. On one hand, a thick film is desirable because it can maximize the number of photons absorbed by the film. On the other hand, increasing film thickness also results in the increase in electrical resistance. Optimization of device performance requires a careful control in device thickness. LbL polymer film deposition is easily to perform, and the waste of materials can be kept at minimum. When a photosensitizing polymer is co-deposited with a charge transport polymer, the polymer molecules essentially form an interpenetrating network (Schlenoff et al., 1998), in which charges can be transported in these network once the excitons are separated.

15.2 Results and Discussion

In this work, the photosensitizing material is based on a ruthenium bisterpyridine complex containing conjugated polymer (polymer 1) (Ng and Chan, 1997), which serves as the polycation. Ruthenium polypyridine complexes have been known to be efficient photosensitizers. The polymer exhibits a strong absorption at 500 nm, which is due to the metal-to-ligand charge transfer transition. The excited states are of triplet in nature, which may facilitate the exciton dissociation process. The anionic charge transport polymer is sulfonated polyaniline (SPAN) (Yue et al., 1991). The structures of these polymers are shown in Fig. 15.1.

---

* Department of Chemistry, The University of Hong Kong, Pokfulam Road, Hong Kong

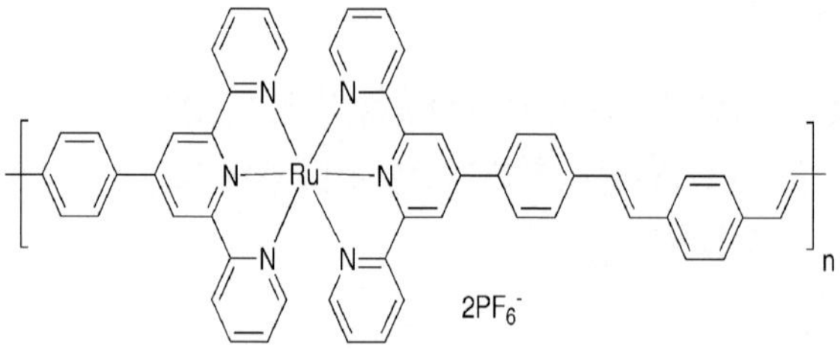

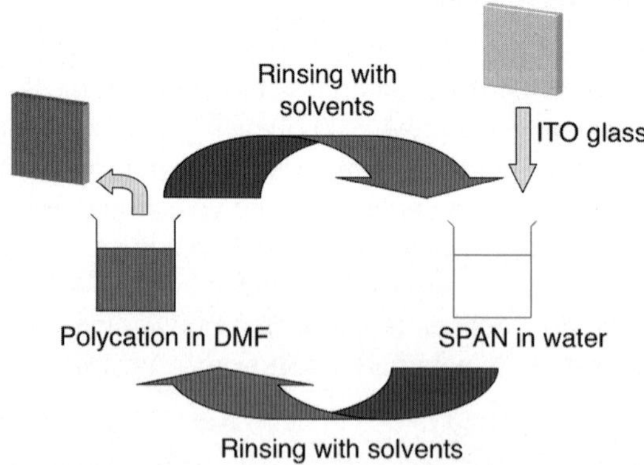

polymer 1

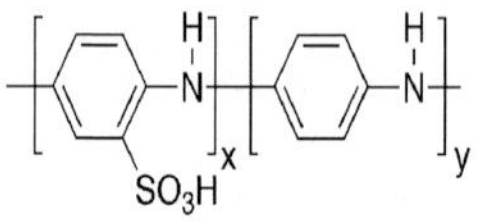

SPAN

Fig. 15.1 Polymers used in device fabrication

Multilayer polymer films were fabricated by immersing the indium-tin-oxide ITO glass substrates in the two polymer solutions in sequence. Since polymer 1 is only soluble in DMF and SPAN is only soluble in water, the substrates have to be immersed in another transition solvent (ethanol) before transferring the substrate from an aqueous to DMF solution in order to prevent precipitation of polymers. The schematic diagram of the film deposition process is shown in Fig. 15.2.

The polymer film deposition conditions were studied in detailed by varying the solution pH, salt concentration, and post-deposition treatment (Decher and Schlenoff, 2003). Multilayer thin films with 20 bilayers were prepared under different conditions. After deposition, all the films were dried under vacuum at room temperature for 1 day. The gradual increase in polymer film thickness was monitored by different methods including optical absorption, quartz crystal microbalance, and spectroscopic ellipsometry.

Table 15.1 summarizes the effect of varying the deposition conditions to the film properties. It can be seen that the rms roughness of the films are not affected by the

Fig. 15.2 Schematic diagram showing the polymer film deposition process

Table 15.1 Effect of deposition conditions

pH of SPAN solution	Electrolyte used	Thickness (nm)	rms roughness (nm)
3	–	58	2.1
3	KI	48	2.2
3	KPF_6	29	2.4
10	–	42	2.5
10	KI	38	2.2
10	KPF_6	9	2.2

pH and presence of electrolyte in the solution. However, the thicknesses of the multilayer films obtained are strongly dependent on the presence of salt in the polymer solutions. In general, those polymer films prepared in salt-free solutions are thicker than those deposited in solutions with electrolytes, and KPF6 yielded polymer films with smallest thickness. This may be related to the conformation change in polymer molecules in different salt solution.

Photovoltaic cells with device structure ITO/(polymer 1/SPAN)$_n$/Al were fabricated and their photocurrent responses under illumination of simulated solar light were studied. When polymer 1 and SPAN was co-deposited into a multilayer thin film, they did not form distinct and stratified layers. Instead, the polymer molecules in a particular layer may disperse up to 3 or 4 layers away from its nominal location within the thin film. The morphology of the film resembles that of a bulk heterojunction photovoltaic cell. The current-voltage characteristics of the device ITO/(polymer 1/SPAN)$_{20}$/Al under illumination with simulated AM1.5 solar light are shown in Fig. 15.3. Compared to the dark current, the photocurrent response increased by 3 orders of magnitude upon light irradiation. The short circuit current density J_{sc}, open circuit voltage V_{oc}, fill factor FF, and power conversion efficiency η_p of the device were measured to be $32\,\mu A/cm^2$, 0.68 V, 0.21, and $4.4 \times 10^{-3}\%$, respectively. Table 15.2 summarizes the photovoltaic performance of the devices fabricated from polymer 1/SPAN multilayer films with different number of bilayers. It can be seen that there is no significant difference in fill factors among the devices with different number of bilayers. The open-circuit voltages decrease

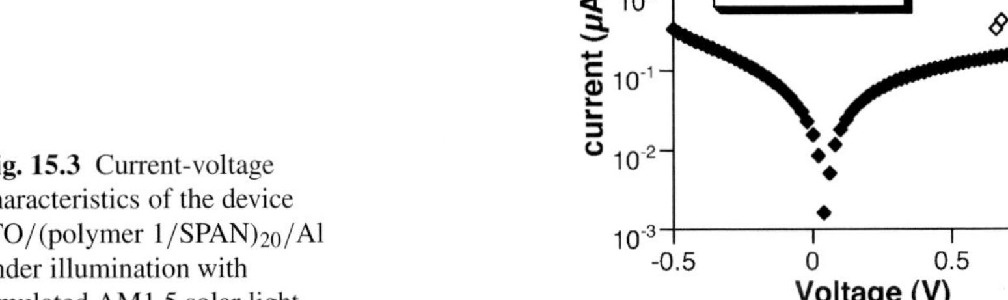

Fig. 15.3 Current-voltage characteristics of the device ITO/(polymer 1/SPAN)$_{20}$/Al under illumination with simulated AM1.5 solar light

Table 15.2 Summary of the photovoltaic device performance

Number of bilayers n	Short circuit current density J_{sc} (μA/cm^2)	Open circuit voltage V_{oc}(V)	Fill factor FF	Power conversion efficiency η_p(10^{-3}%)
20	32.9	0.68	0.21	4.4
40	11.4	0.54	0.22	1.4
60	7.5	0.54	0.24	0.95
40	0.72	0.36	0.21	0.56

slightly with increasing the number of bilayers, which might be due to the exciton recombination. For the device with 20 bilayers, it exhibits the best performance. Compared to the device with thicker active layer, increasing the multilayer thickness results in enhancement in optical absorption, the electrical resistance of the film also increased.

In order to study the role of photosensitizers in photovoltaic process, we measured the incident photon to electron conversion efficiency (IPCE) of the devices at different wavelength. IPCE is the number of electron generated by the device per photon absorbed, which is defined as:

$$IPCE\ (\%) = \frac{1240 \times J_\lambda (A/cm^2)}{P_{in}(W/cm^2) \times \lambda(nm)} \times 100\% \tag{15.1}$$

where J_l is the current density, P_{in} is the incident light power, and l is the wavelength of the incident light. From the IPCE vs. l plot, we are able to study which functional group in the polymer is responsible for the sensitization process. This provides important information in modifying the polymer structure for device optimization. Figure 15.4 shows the plot of IPCE vs. wavelength and the absorption spectrum of the multilayer film for the device ITO/(polymer 1/SPAN)$_{20}$/Al. The peak observed at *ca.* 500 nm clearly demonstrates that the generation of photocurrent is mainly contributed by the ruthenium complex in the polymer.

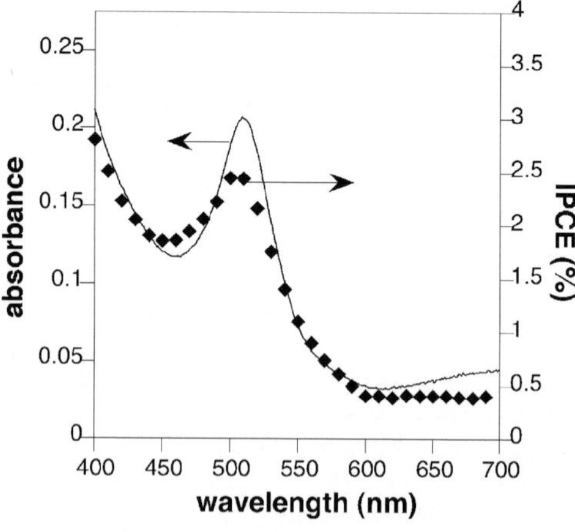

Fig. 15.4 Plot of IPCE as the function of incident light wavelength and the absorption spectrum for the device ITO/(polymer 1/SPAN)$_{20}$/Al

15.3 Conclusions

A series of photovoltaic cells were fabricated by the layer-by-layer deposition process using a ruthenium containing polymer and sulfonated polyaniline. This is a versatile approach in preparing polymer thin film devices because the film thickness can be accurately controlled. It was found that the film properties are dependent on the fabrication conditions, and the generation of photocurrent was mainly due to the ruthenium terpyridine complexes. These devices showed photocurrent response upon irradiation of simulated solar light, and the power conversion efficiencies are in the order of $10^{-3}\%$. The low efficiency measured may be due to the low optical absorbance and exciton recombination in the films. In addition, the effect of the presence of electrolytes in the films is not well understood. More detailed work will be required in order to understand the fundamental mechanisms in the photosensitization, exciton separation, and charge transport processes in the film.

Acknowledgments The work is substantially supported by the Research Council of The Hong Kong Special Administrative Region, China (Project Nos. HKU 7009/03P, 7008/04P, and RGC Central Allocation HKU2/05C). Financial supports from the Committee on Research and Conference Grant (University of Hong Kong) are also acknowledged.

References

Decher, G. and Hong. J. D., 1991, Buildup of ultrathin multilayer films by a self-assembly process. 1. Consecutive adsorption of anionic and cationic bipolar amphiphiles on charged surfaces. *Makromolekulare Chemie-Macromolecular Symposia* **46**, pp. 321–327.

Decher, G. and Schlenoff, J. B., 2003, *Multilayer Thin Films: Sequential Assembly of Nanocomposite Materials* (Weinheim: Wiley-VCH).

Ng, W. Y. and Chan, W. K., 1997, Synthesis and photoconducting properties of poly(p-phenylenevinylene) containing a bis(terpyridine) ruthenium(II) complex. *Advanced Materials* **9**, pp. 716–719.

Schlenoff, J. B., Ly, H., Li, M., 1998, Charge and mass balance in polyelectrolyte multilayers. *Journal of The American Chemical Society* **120**, pp. 7626–7634.

Yue, J., Wang, Z. H. et al., 1991, Effect of Sulfonic Acid Group on Polyaniline Backbone. *Journal of The American Chemical Society* **113**, pp. 2665–2671.

Chapter 16
Optical Properties of Arrays of Iodine Molecular Chains Formed Inside the Channels of AlPO$_4$-5 Zeolite Crystals

J. T. Ye and Z. K. Tang[*]

16.1 Ordered Nanostructures Synthesized Using Zeolite Frameworks

Condensing guest molecular species in the nano pores of zeolite from the vapor phase form array of nanostructures which replicated the periodicity of the zeolite frameworks. By selecting a proper combination of host framework and guest material, one can realize various nano structures with different ordered spatial assembling which exhibit a variety of interesting characteristics (Tang et al., 1998; Koide et al., 2001; Kurbitz et al., 2001; Li et al., 2001; Toshima et al., 2001). Examples include semiconductor nano clusters and wires (Bogomolov, 1973, 1976; Goto et al., 1993; Poborchii et al., 1996; Nozue et al., 1996; Poborchii, 2001), organic zeolite lasers (Vietze et al., 1998), and zeolite-based second harmonic generators (Herance et al., 2004).

16.2 Fabrication of Iodine Molecular Chains Inside AlPO$_4$-5 Zeolite

Iodine is the heaviest common halogen (atomic number: 53, atomic mass: 127) exists as a solid at room temperature in sublimation equilibrium with its vapor. Like other halogens in their solid phase, the iodine crystal consists of weakly bound diatomic molecules, I_2. Iodine sublimates even at low temperature and generates enough vapor pressure facilitating its insertion and diffusion into the zeolite channels which act as a natural template to guide the growth of iodine nanostructures.

Figure 16.1a shows the method of condensing iodine molecules to the AlPO$_4$-5 (IUPAC code: AFI) zeolite framework schematically. Pure AFI crystals are obtained by calcination the tripropylamine $(CH_3CH_2CH_2)_3N$ organic templates at 580 °C in an O_2 atmosphere (Qiu et al., 1989; Tang et al., 1997; Sun et al., 1999). To prepare

---

* Department of Physics and Institute of Nano-Science and Technology, Hong Kong University of Science and Technology, Clear Water Bay, Kowloon, Hong Kong, China
e-mail: phzktang@ust.hk

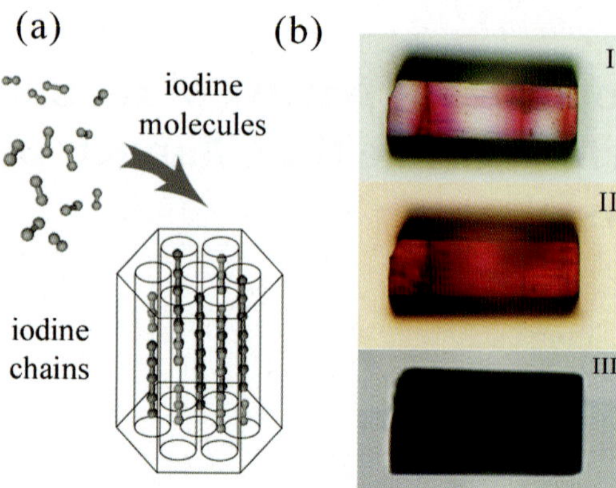

Fig. 16.1 (a) Schematic presentation of formation of iodine chains from iodine molecules. (b) The diffusion process of occluding iodine molecules into the channels of AFI zeolite crystals. (I), Iodine molecules colour the opened ends of the zeolite crystal at room temperature. (II) Iodine molecules diffuse along the channel and color the whole crystal at around 100 °C. (III) High density iodine structures are formed inside the channel after heating the zeolite crystal in the iodine vapor at 400 °C for several hours

the iodine structures in the nano pores of the AFI crystals (iodine@AFI), calcinated AFI zeolite is dehydrated at 580 °C for 2 hours under a vacuum, and then mixed with solid iodine source (BDH 99%) in a Pyrex tube inside a moisture-free ($<$ 1 ppm) glove box. The tube is then sealed under a vacuum of around 1×10^{-1} m bar.

As shown in the Fig. 16.1b, even at room temperature, iodine molecules enter the channels of $AlPO_4$-5 zeolite crystals as evidenced by the red-brown colour at the opened ends of the zeolite crystal. The opened ends are either the two natural facets of the crystal or the cracks formed during the calcination process. By increasing temperature to around 100 °C, iodine molecules diffused along the channels and distributed to the whole crystal as shown in stage II. While in stage III, high density of iodine structures were occluded inside the channels of the AFI crystal after heating the crystal at around 400 °C for several hours. The iodine@AFI crystal is optically isotropic when the iodine density is relatively low inside the channel, while optical anisotropy is observed when the density becomes higher (Ye et al., 2006). As shown in stage III, the iodine@AFI crystal shows a uniform dark purplish color, indicating a higher iodine density.

16.3 Optical Characterization of Iodine Chains

In this section, optical measurements of the ordered one-dimensional array of stable iodine molecular chains formed by iodine molecules adsorbed inside the channels of AFI zeolite are discussed. Polarized Raman scattering showed that these chains were almost perfectly aligned to the c-axis of the AFI crystal. Two chain species, I_n^- and $(I_2)_n$, were distinguished by measuring polarized optical absorption and resonant Raman scattering.

16.3.1 Raman Scattering From iodine@AFI with Different Densities

The different loadings of the iodine molecules inside AFI zeolite were characterized by means of Raman scattering excited using the 514.5 nm line of an Ar/Kr ion laser. The photon energy of this laser line corresponds to a transition from the lowest vibrational levels of the singlet electronic ground state to high vibrational levels of a triplet excited state. In the low diffusion stage, the Raman spectrum of the iodine species inside the zeolite channels exhibited a typical spectral feature of diatomic iodine molecule in the vapor phase as shown in the upper panel of Fig. 16.2. This Raman spectrum is characterized by the principal Raman signal at $212 \, \text{cm}^{-1}$ with $\Delta v = 1$, and the other features at $419, 625, 829, 1032 \, \text{cm}^{-1}, \ldots$ correspond to vibrational energy levels with $\Delta v = 2, 3, \ldots$, respectively. The principle vibrational frequency of vapour phase iodine molecules is at $214 \, \text{cm}^{-1}$. The slight downshift of $2 \, \text{cm}^{-1}$ can be explained by the increased interaction between the iodine species and the zeolite frameworks (Wirnsberger et al., 1997, 1999; Hertzsch et al., 2002). The Raman spectrum of low density iodine indicates that the iodine species incorporated in the channels are molecular vapor. The structures formed inside the zeolite crystals are very sensitive to the diffusion temperature. With increase of the diffusing temperature, higher density of iodine molecules can be loaded into the zeolite channels. At high loading density, two new Raman features at 110 and $175 \, \text{cm}^{-1}$, are observed at the low frequency side of the molecular principle vibration, indicating new iodine structures are formed inside the AFI channels. From stage I to III, the molecular vibration frequency shifted from 212 to $206 \, \text{cm}^{-1}$, due to the increasing interaction between the iodine molecules at high loading density. As a result of greater interaction, the vibrational frequencies of Raman modes emerged in stage II are also downshifted from 175 to $168 \, \text{cm}^{-1}$ with the increase of iodine density from stage II to III.

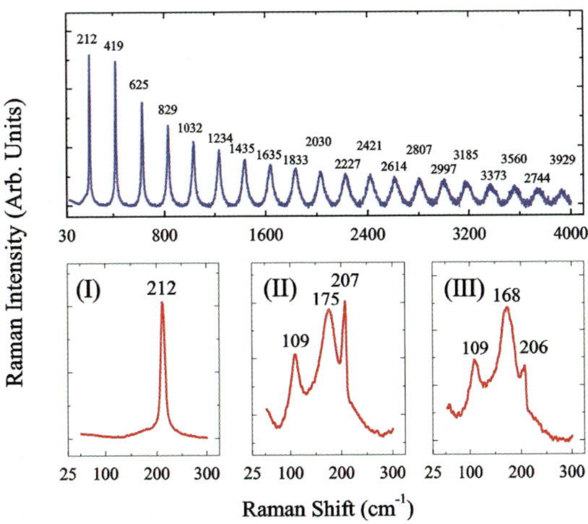

Fig. 16.2 The upper penal shows the Raman spectrum of an iodine@AFI crystal with a low iodine loading. The lower panel shows the evolution of the Raman spectra when the iodine density increases from stage I to III as shown in Fig. 16.1

16.4 Polarized Raman Spectroscopy Characterization

Polarized Raman spectra of the iodine@AFI single crystals were measured on a micro-Raman system (Jobin Yvon T64000) in a backscattering configuration using the 514.5 nm excitation line from an Ar/Kr-ion laser. As shown in Fig. 16.3, for both *VH* (polarization of the excitation laser light and the detection of Raman scattering are perpendicular) and *VV* (polarization of the excitation laser light and the detection of Raman scattering are parallel) configurations, highly polarized Raman modes at 110 and 168 cm^{-1} were observed in the stage III samples. In the *VH* configuration, the intensity of these two modes increases from $\theta = 0°$ to $\theta = 45°$ and decreases from $\theta = 45°$ to $\theta = 90°$ with a maximum at $\theta = 45°$. In the *VV* configuration, the intensity decreases dramatically from $\theta = 0°$ to $\theta = 90°$ with a maximum at $\theta = 0°$. Obvious anisotropy of the iodine@AFI system in the polarized Raman scattering indicates that these two anisotropic modes are associated with one-dimensional structures formed inside the channels. The vibration mode at 110 cm^{-1} can be attributed to linear I_3^- chains, which were previously observed at a similar frequency in iodine solution (Kiefer and Bernstein, 1972), defects in CsCl (Martin, 1976), as well as liquid iodine (Magana and Lannin, 1985). The formation of ionic chains can originate from the charge transfer between iodine chains and crystal defects or residual carbonaceous clusters inside the channels. The Raman mode at 168 cm^{-1} is originated from long $(I_2)_n$ chains that have been observed at 175–185 cm^{-1} in liquid iodine as well as amorphous iodine (Shanabrook et al., 1981; Magana and Lannin, 1985). It is worth noting that the vibration frequency of long $(I_2)_n$ chains inside the AFI channels is substantially lower than that observed in liquid iodine. The red shift in the frequency is due to longer and more stable chain structure formed inside the AFI

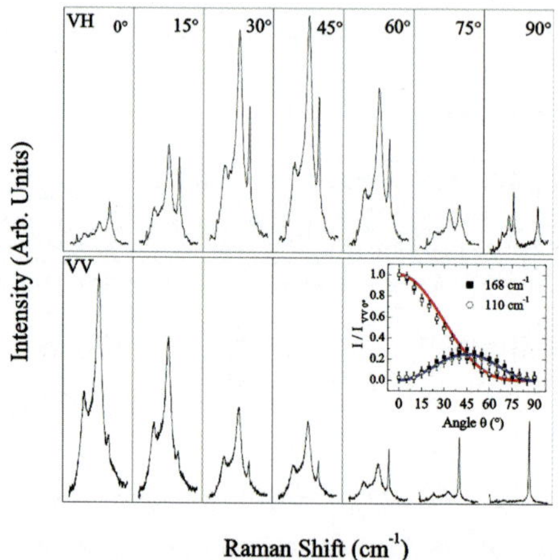

Fig. 16.3 Polarized Raman spectra of iodine@AFI at *VH* and *VV* configurations are shown in the top and bottom panels, respectively. The inset shows the angle dependence of Raman intensities in the *VH* and *VV* configurations. The relative intensities of $I_{VH}/I_{VV0°}$ and $I_{VV}/I_{VV0°}$ are fitted as functions of $\cos^2\theta \cdot \sin^2\theta$ and $\cos^4\theta$, respectively

channels. The spectral width of the Raman band at $168\,cm^{-1}$ is broad, suggesting that this mode is composed of Raman scattering from $(I_2)_n$ chains of many different lengths. It is also possible to contain contributions from species like I_n^- with $n>3$ vibrating at frequencies in a similar range of the Raman spectrum (Mulazzi et al., 1981). The series of intra-molecular vibrations observed at $209\,cm^{-1}$ is attributed to vapor-like molecular species. In iodine vapor, these modes are usually observed at $214\,cm^{-1}$. The slight $5\,cm^{-1}$ downshift can be explained by a stronger inter-action between iodine molecules inside the zeolite channels (Wirnsberger et al., 1997, 1999; Hertzsch et al., 2002). The vapor-like molecular iodine is expected in a random orientation as its Raman intensity at $209\,cm^{-1}$ shows no polariza-tion dependence. Since the 6.8 Å Van der Waals radius of molecular iodine is slightly smaller than the 7.3 Å channel diameter of AFI, it is geometrically allow-able for iodine molecules to rotate inside the channels. However, species like I_n^- and $(I_2)_n$ with chains composed of more than three iodine atoms can only align to the channel direction since a three-iodine linear chain is already wider than the channel.

Micro polarized absorption is measured on a single crystal of iodine@AFI which is mechanically polished into a $10\,\mu m$ thick foil. As shown in Fig. 16.4a, the optical absorption spectra have strong polarization dependence with two absorption bands at 2.3 and 2.7 eV, respectively, exhibiting high absorption for the light polarized parallel to the channel direction and high transparency for the light polarized perpendicular to the channel direction. Another absorption band, which is not sensitive to polarization, is observed at 2.38 eV. This absorption band can be attributed to the molecular iodine inside the channels, as its absorption energy coincides well with the $B \leftarrow X$ transition (2.39 eV) of vapor phase iodine molecules (Mulliken, 1971). It is intuitive to attribute the polarized absorption bands to the chain structures. These iodine chains are isolated by the channel walls and they are almost perfectly aligned along the channel direction. This configuration of one-dimensional arrays has an inherent strong depolarization effect (Ajiki and Ando, 1994). The contribution of the anisotropic bands is significant when the polarization angle is relatively small, while it is negligible for the spectrum at $\theta = 90°$. In Fig 16.4a, the spectra at $\theta < 75°$ are sub-tracted from the spectrum at $\theta = 90°$ to separate the polarized and unpolarized bands since the spectrum at $\theta = 90°$ shows only absorption from molecular iodine.

16.5 Resonant Raman Scattering and Micro Optical Absorption

Resonant Raman scattering is utilized to distinguish the origin of the anisotropic absorption bands. As shown in the left panel of Fig.16.4b, Raman scattering is excited using the laser lines with wavelength ranging from 457.9 to 647.1 nm. The Raman bands of the iodine chains at 110 and $168\,cm^{-1}$ exhibit obvious resonant behaviour. For Raman signals of the I_3^- and $(I_2)_n$ chains, the resonant maxima are observed at the laser excitations of 2.66 and 2.34 eV, respectively. As shown by

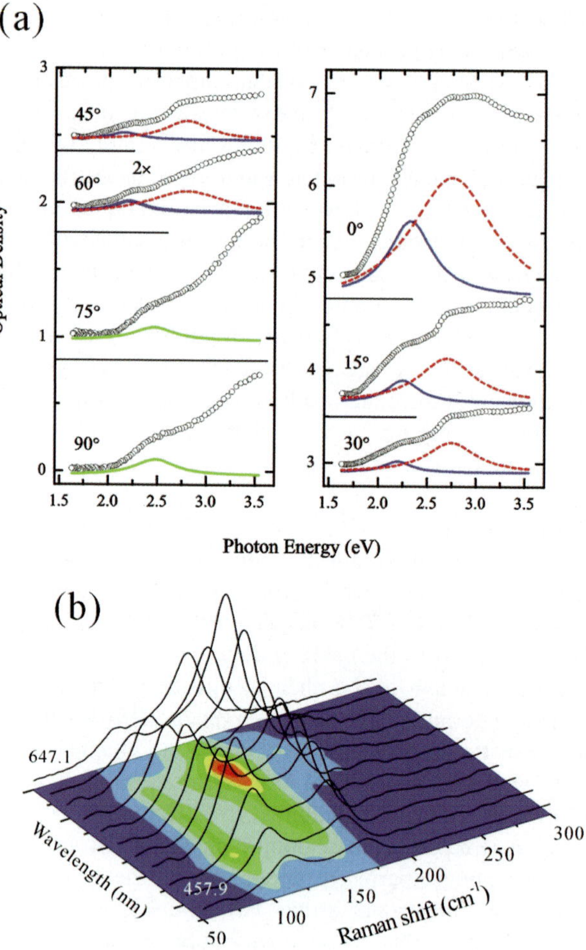

Fig. 16.4 (a) Polarized optical absorption spectra of an iodine@AFI crystal. Spectra labeled $\theta = 0°$ and $\theta = 90°$ stand for measurement when the polarization of incident light are parallel and perpendicular to the crystal axis, respectively. (b) Normalized resonant Raman spectra of iodine@AFI with excitation wavelengths from 457.9 to 647.1 nm. Resonant maxima of 110 and 168 cm^{-1} bands are observed at 465.8 and 530.9 nm, respectively

the intensity mapping at the bottom of Fig.16.4b, the profile of normalized Raman intensity of these two bands versus the excitation energies indicates resonant Raman intensities of I_3^- and $(I_2)_n$ chains coincide with the optical absorption band at 2.3 and 2.7 eV, respectively. The coincidence of the resonant laser energy with the energy of the absorption band suggests that the optical absorption at 2.7 eV can be assigned to the I_3^- chains and 2.3 eV to the $(I_2)_n$ chains. The anisotropic absorption band at around 2.3 eV (540 nm) is related to the $B \leftarrow X$ (518 nm) transition in vapor phase iodine that shows an evident red shift (22 nm) due to the formation of $(I_2)_n$ chains from individual I_2 molecules. The absorption band of I_3^- corresponding to the $^1\Sigma \rightarrow {}^1\Pi$ transition is at around 2.6 eV (Mulazzi et al., 1981), which is close to the anisotropic absorption band observed at 2.7 eV.

In summery, a uniform array of one-dimensional molecular iodine chains is formed inside the AFI zeolite frameworks using vapor phase diffusion method. The structures of iodine inside the AFI channels are highly dependent on the density of iodine inside the channels characterized by different Raman scattering for iodine densities at different diffusion stages. Polarization dependence of intensity of Raman modes of iodine species indicates that the chains are aligned perfectly along the c-axis of AFI crystal. Polarized optical absorption as well as resonant Raman scattering reveal two chain species inside the channels of AFI zeolite: I_n^- and $(I_2)_n$.

Acknowledgments This research was supported by Hong Kong CERGE Grants of 602807, RGC DAG 05/06.sc33, RPC 06/07.sc06, and HKUST President direct allocation F0204-A.

References

Ajiki, H. and T. Ando, 1994, *Physica B* **201**, p. 349.

Bogomolov, V. N. 1973, *Fiz. Tverd. Tela Leningrad* **15**, p. 1312.

Bogomolov, V. N. 1976, *Pis'ma Zh. Eksp. Teor. Fiz.* **23**, p. 528.

Goto, T., Y. Nozue, et al., 1993, *Mater. Sci. Eng. B* **19**, p. 48.

Herance, J., D. Das, et al., 2004, *Chem. Phys. Lett.* **395**, p. 186.

Hertzsch, T., F. Budde, et al., 2002, *Angew. Chem. Int. Ed.* **41**, p. 2282.

Kiefer, W. and H. J. Bernstein, 1972, *Chem. Phys. Lett.* **16**, p. 5.

Koide, T., H. Miyauchi, et al., 2001, *Phys. Rev. Lett.* **87**, p. 257201.

Kurbitz, S., J. Porstendorfer, et al., 2001, *Appl. Phys. B* **73**, p. 333.

Li, Z. M., Z. K. Tang, et al., 2001, *Phys. Rev. Lett.* **87**, p. 127401.

Magana, R. J. and J. S. Lannin, 1985, *Phy. Rev. B* **32**, p. 3819.

Martin, T. P., 1976, *Phys. Rev. B* **13**, p. 3618.

Mulazzi, E., I. Pollini, et al., 1981, *Phy. Rev. B* **24**, p. 3555.

Mulliken, R. S., 1971, *J. Chem. Phys.* **55**, p. 288.

Nozue, Y., T. Kodaira, et al., 1996, Magnetic Properites of Alkali Metal Cluster in Zeolite Cages. *Mater. Sci. Eng. A* **217–218**, p. 123–128.

Poborchii, V. V., 2001, Raman microprobe polarization measurement as a tool for studying the structure and orientation of molecules and clusters incorporated into cubic zeolites: S8 and S12 rings in zeolite A. *J. Chem. Phys.* **14**, p. 2707.

Poborchii, V. V., M. S. Ivannova, et al., 1996, Raman and absorption spectra of the zeolites A and X containing selenium and tellurium in the nanopores. *Mater. Sci. Eng. A* **217–218**, pp. 129–134.

Qiu, S., W. Pang, et al., 1989, *Zeolites* **9**, p. 440.

Shanabrook, B. V., J. S. Lannin, et al., 1981, *Phys. Rev. Lett.* **46**, p. 130.

Sun, H. D., Z. K. Tang, et al., 1999, *Solid State Commun.* **109**, p. 365.

Tang, Z. K., M. M. T. Loy, et al., 1997, *Appl. Phys. Lett.* **70**, p. 34.

Tang, Z. K., H. D. Sun, et al., 1998, *Appl. Phys. Lett.* **73**, p. 2287.

Toshima, N., Y. Shiraishi, et al., 2001, *J. Mol. Catal.* **177**, p. 139.

Vietze, U., O. Krasb, et al., 1998, Zeolite-Dye Microlaser. *Phys. Rev. Lett.* **81**, p. 4628.

Wirnsberger, G., H. P. Fritzer, et al., 1997, *J. Mol. Struct.* **410**, p. 123.

Wirnsberger, G., H. P. Fritzer, et al., 1999, *J. Mol. Struct.* **481**, p. 699.

Ye, J. T., Z. K. Tang, et al., 2006, *Applied Physics Letters* 88, pp. 073114.1–073114.3.

Part IV
Molecular Electronics

sity of single-molecule manipulations using the STM tip to create novel artificial nanostructures mechanically (Stroscio and Eiger, 1991; Eigler and Schweizer, 1990; Neu et al., 1995; Jung et al., 1996) or chemically (Hla and Rieder, 2003; Ho, 2002; Lorente et al., 2005; Hou and Zhao, 2006; Hou et al., 2004). In this paper we report the quantum manipulation and the realization of some nanodevices at the molecule scale with STM.

17.2 Manipulating the Electronic Properties of Single Molecules by Chemical Doping

Fullerene molecules have attracted much attention in recent years due to their unique properties in both bulk and single-molecule states. They are good candidate for realizing nanometer molecular devices due to their nanometer structures and distinct electronic properties. Chemical doping of fullerenes will greatly change their electronic properties which can be investigated with STM. According to the principles of electron tunneling, in the simplest case, the tunneling current is proportional to a convolution of the DOS of the STM tip and the DOS of the sample over an energy range eV, where V is the bias voltage applied (Chen, 1993). Therefore, if assuming tip DOS is constant, the dI/dV spectroscopy reflects the LDOS of the sample near E_F (Lu et al., 2003b). The dI/dV technique can be used for investigating complex electronic structures of individual molecules especially those can not be detected by common STM measurements. We showed that the energy-resolved metal-cage hybrid states of a single endohedrally doped metallofullerene $Dy@C_{82}$ isomer I adsorbed on a $Si(111) - \sqrt{3} \times \sqrt{3} - Ag$ surface can be spatially mapped via STM dI/dV technique (Wang et al., 2003b), supporting a complex picture consisting of the orbital hybridization and charge transfer for the interaction between the cage and the metal atom.

In most cases, the electronic properties of single molecules adsorbed on metal or semiconductor surfaces are strongly influenced by the substrate electrons. By introducing another tunneling barrier, usually a thiol monolayer or a thin oxide film, to isolate single molecules and forming a double-barrier tunneling junction (DBTJ) configuration, the STM can take advantage of some unique and intrinsic properties of the molecules to realize electronic devices with specific functions. We found that the azafullerene $C_{59}N$ molecule, a kind of chemically doped C_{60}, can be used as a rectifier in a STM-based DBTJ via the single electron tunneling (SET) effect (Zhao et al., 2005b). A fullerene molecule in a DBTJ show SET effect analogous to semiconductor quantum dot with zero-current Coulomb Blockade (CB) around the Fermi surface (Li et al., 2006). For fullerene C_{60} with a full-occupied highest occupied molecular orbital (HOMO) and lowest unoccupied molecular orbital (LUMO), we observed a symmetrically CB behavior in the I-V curve where the current onsets at both negative and positive sample bias voltage are approximately equal. While, for azafullerene $C_{59}N$ with a half-occupied HOMO, pronounced I-V asymmetry was observed and the rectifying effect exists in all measurements we

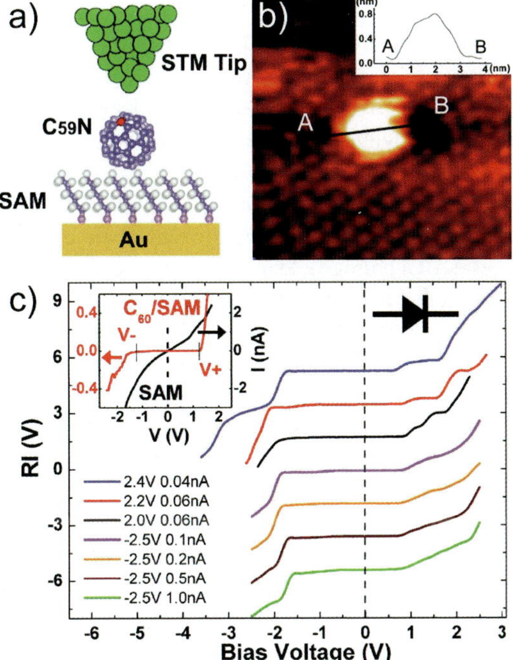

Fig. 17.1 (a) The diagram of the experimental system. (b) The STM image of an isolated C59N adsorbed on thiol SAM at 5 K. Image size: 8 nm × 8 nm, scanning parameters: 1.86 V/0.1 nA. Inset shows line profile along the line AB. (c) A set of *I-V* curves for individual C59N molecules measured at 5 K and at various setting parameters. The inset shows the *I-V* curves for the SAM substrate and individual C_{60}. (From Zhao et al., 2005b, Copyright 2000 by the American Institute of Physics)

made. The positive onset sample bias voltage was always about 0.5–0.7 V, while the negative onset voltage was about 1.6–1.8 V (Fig. 17.1). By theoretical analysis we found that the half-occupied HOMO and the asymmetric shift of the Fermi level when the molecule was charged (the HOMO-LUMO gaps of $C_{59}N^-$ and $C_{59}N^+$ are distinctly different) were responsible for the molecular rectification.

17.3 Controlling the Transport Properties of Single Molecules and Nanoparticles by Direct Manipulation

Direct manipulation including vertical manipulation and lateral manipulation of single adsorbates can be used to construct novel tunneling structures such as an NDR device involving two fullerene molecules vertically coupled. A single fullerene molecule can be accommodated into an STM-based DBTJ system with thiol SAM and manifest as an NDR device (Zeng et al., 2000) Atomic scale NDR effect could arise in the tunneling structures with an STM tip and a localized surface site when the LDOS of both the tip and the surface site have adequately narrow features.

Single C_{60} molecules are great candidates for realize NDR devices in principle because neutral C_{60} molecule has sharp LDOS features owing to its degenerated and well-separated energy states. In a DBTJ system, C_{60} molecules adsorbed on a thiol SAM remain their native properties such as LDOS in the main. A tip with distinct and narrow LDOS features was prepared by controllably attaching a single C_{60} molecule onto its apex. By using such a tip we observed obvious NDR effect on individual C_{60} molecules on SAM (Zeng et al., 2000) (Fig. 17.2).

Nanoparticles are also good candidates for nanoscale electronics devices. The I-V transport properties of a nanparticle in a STM-based DBTJ system will show CB and Coulomb staircases due to its discrete energy levels and the charging effect (Hou et al., 2001; Wang et al., 2001, 2003a, 2004). Metal or semiconductor nanoparticles with different sizes and properties can be readily controlled by chemical methods (Lu et al., 2003a) and the chemical properties of the nanoparticles will greatly affect the quantum transport properties in an STM. For example, size dependent multi-peak spectral features in the differential conductance curve are observed for the crystalline Pd nanoparticles but not for the amorphous particles (Hou et al., 2003) (Fig. 17.3a–c). Theoretical analysis shows that these spectral features are related to the quantized electronic states in the crystalline Pd particle. The suppression of the quantum confinement effect in the amorphous particle arises from the reduction

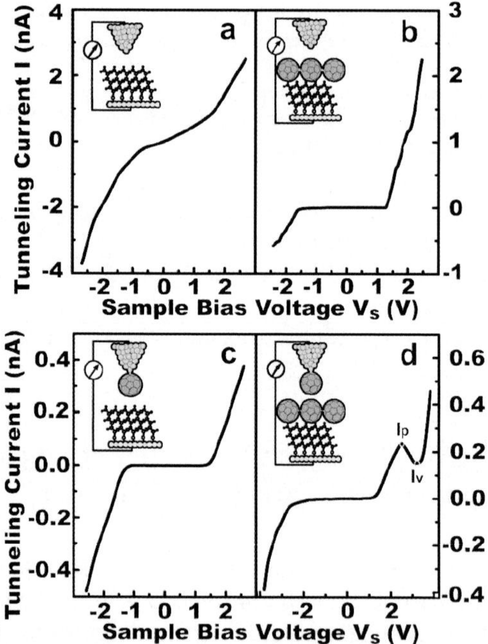

Fig. 17.2 I-V curves obtained from four kinds of tunneling structures, as shown in the insets. (a), (b), (c), and (d) correspond to bare Pt-Ir tip over thiols, bare Pt-Ir tip over C_{60}, C_{60}-modified tip over thiols, and C_{60}-modified tip over C_{60}, respectively. Clear NDR effect was found in (d). (From Zeng et al., 2000, Copyright 2000 by the American Institute of Physics)

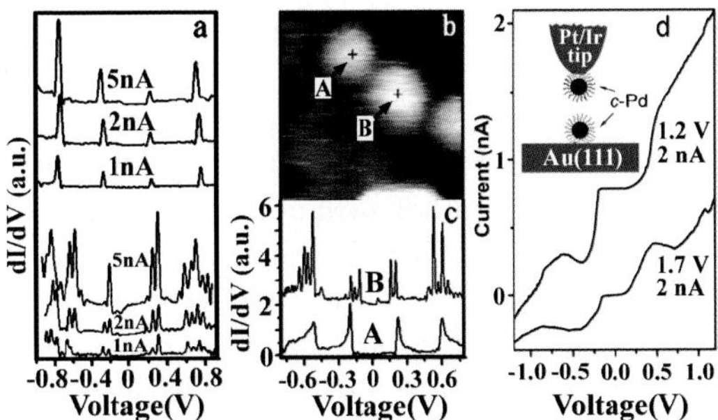

Fig. 17.3 (a) The numerical differential conductance spectra of two 2 nm amorphous Pd and crystalline Pd (c-Pd) particles with various set point of the tunneling currents. (b) A STM image showing an amorphous Pd particle A and a c-Pd particle B; (c) dI/dV spectra acquired on particle A and particle B, respectively. (d) I-V curves measured on two coupled c-Pd particles with the structure depicted as inset. The zero current gaps of curves were centered at zero bias for comparison. (From Hou et al., 2003, Copyright 2001 by the American Physical Society)

of the degeneracy of the eigenstates and the level broadening due to the reduced lifetime of the electronic states.

Due to the narrow features of the discrete energy levels, similar to fullerene molecules, nanoparticles can also be used to realize NDR devices (Hou et al., 2003) by directly manipulating a crystalline Pd nanoparticle to the STM tip (Fig. 17.3d).

17.4 Manipulating the Magnetic Properties of Adsorbed Ions by Single-molecule Chemistry

Scanning tunneling microscope can be employed to perform single-molecule chemistry by injecting energetic tunneling electrons into individual adsorbed molecules and hence selectively exciting a specific chemical bond or a vibrational mode of the molecule. These kinds of electron induced manipulations provide more powerful abilities in single-molecule chemical modification and allow formation of novel artificial molecular structures and hence tuning of the properties of the molecules.

Ligand molecules like phthalocyanines and porphrins serve as good carriers for magnetic metal ions. Using an STM, we show that the Kondo effect of the magnetic moment of a single adsorbed ion can be controlled by altering its chemical environment (Zhao et al., 2005a). The Kondo effect arises from the coupling between localized spins and conduction electrons. At sufficiently low temperatures, the coupling can lead to changes in the transport properties through scattering or resonance effects (Kondo, 1964; Kouwenhoven and Glazman, 2001) which are signatures of the presence of magnetism.

For a cobalt phthalocyanine(CoPc) molecule, a magnetic Co^{2+} ion is embedded in its center, but the magnetism totally quenched when the molecule is adsorbed onto the Au surface due to the electronic interactions between the Co ion and the substrate. The nonmagnetic behavior can be verified both from experimental dI/dV spectroscopy and from theoretical calculations. However, after cutting away eight hydrogen atoms from the molecule with high-energy tunneling electrons emitted from a STM tip, both the electronic and geometric structures of the molecule were changed remarkably. The feasibility of such a manipulation was previous demonstrated in the dehydrogenation experiments of single benzene molecules on copper surfaces (Komeda et al., 2004; Lauhon and Ho, 2000) in which a benzene bonding geometry change from flat-lying to up-right configuration was observed and the breaking of C-H bonds were proposed to be induced by the formation of the transient negative ion due to the trapping of the injected electrons at the unoccupied π^* resonant states of the benzene molecules. In our experiment, when we apply a positive high voltage (>3.3 V) over the edge of a lobe, the I-t curve shows two sudden drops indicating the sequential dissociation of the two end H atoms from the benzene ring in the lobe (Fig. 17.4). After the dehydrogenation process, the lobe disappeared in topographic image, indicating a strong chemical bonding of the highly-reactive benzene ring to the Au surface. After pruning off all eight hydrogen atoms from the periphery of the CoPc molecule, all the four lobes disappeared and the molecule arched away from the surface. In this fully-dehydrogenated CoPc (d-CoPc) molecule, we found a strong resonance at E_F arised in the dI/dV spectroscopy, however, the original d_z^2-orbital induced resonant state located at $-150\,mV$ in a pristine CoPc molecule now completely disappeared (Fig. 17.5). This zero-bias resonance was attributed to the Kondo effect associated with Fano resonance by careful theoretical analysis. First-principles calculations and simulations showed the cobalt atom in a free CoPc has unpaired d electrons and the

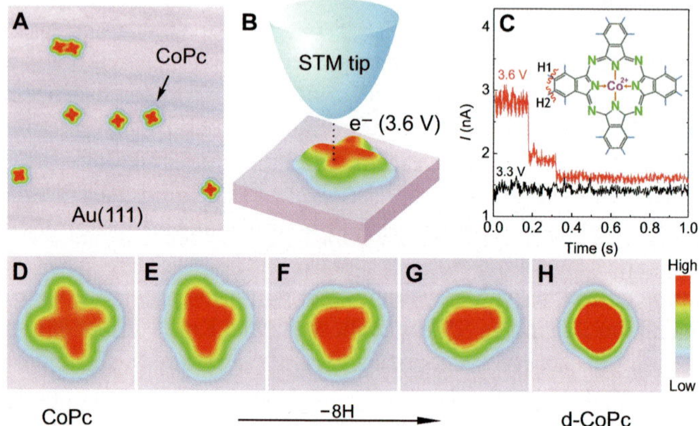

Fig. 17.4 (A) individual CoPc molecules adsorbed on Au(111) surface. (B) Diagram of the dehydrogenation induced by the tunneling electrons. (C) Current traces during two different voltage pulses on the brink of one lobe. Inset shows the molecular structure of CoPc and the two hydrogen atoms to be pruned off in each one lobe. (D to H) STM images (25Å by 25 Å, V = 1.2 V, I = 0.4 nA) of a single CoPc molecule during each step of the dehydrogenation process, from (D) an intact CoPc to (H) a final dehydrogenated CoPc (d-CoPc). The color scale represents apparent heights, ranging from 0 Å (low) to 2.7 Å (high)

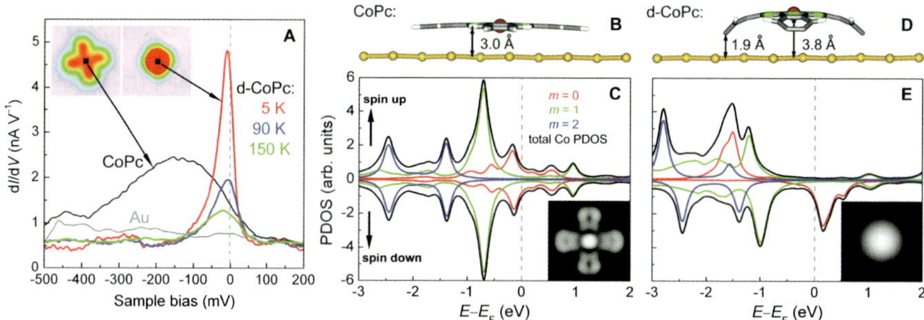

Fig. 17.5 (A) dI/dV spectra measured directly over center Co atoms of CoPc and d-CoPc. All spectra were taken with the same set point of $V = 600\,\text{mV}$ and $I = 0.4\,\text{nA}$. (B and D) Side views of the optimized structures for CoPc/Au(111) and d-CoPc/Au(111) adsorption systems, respectively. (C and E) The PDOS of the Co atom in a CoPc and a d-CoPc molecule on an Au(111) surface, respectively. Insets are the corresponding simulated STM images

magnetic moment is 1.09 Bohr magnetons (μB). But in the CoPc/Au(111) adsorption system, the magnet moment of the Co atom is completely quenched by the molecule-substrate interaction. In the d-CoPc/Au(111) artificial molecular structure, the magnet moment of the Co atom is recovered to be 1.03 μB, very close to the value of a free CoPc molecule. This recovered localized spin in the Co atom as well as the chemical bonding of the d-CoPc ligand lead to the formation of Kondo ground state at low temperatures and result in the observed Kondo effect.

These experiments of electron-excited single-molecule manipulation demonstrate the feasibility of an STM in manipulating not only the electronic properties but also the magnetic properties of a single molecule that promises potential applications in future molecular electronics and spintronics on the atomic level.

17.5 Summary

In conclusion, we try to manipulate the electronic structures and the transport properties of single molecules and nanoparticles by controlled doping, direct manipulation, and single-molecule chemistry. By introducing single molecules or nanoparticles into the tunnel junctions in which the tip of STM and supporting surface are two electrodes, quantum transport properties were studied and their electronic structures were determined from the results of STM combined with the theoretical modeling. Novel effects associated with the discrete energy levels and the charging energy of ultra small tunneling junction, like NDR effect, rectifying and Kondo effect at the molecular scale are reported.

Acknowledgments This work is partially supported by the Ministry of Science and Technology of China (grant nos. G1999075305, G2001CB3095, and G2003AA302660) and the National Natural Science Foundation of China.

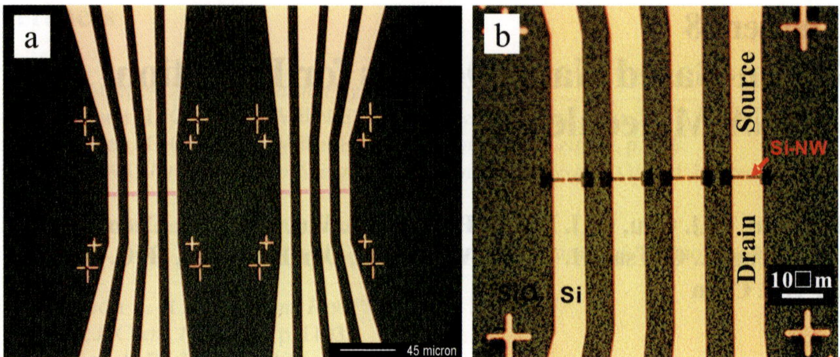

Fig. 18.1 Optical micrographs of the SOI-based silicon nanowire devices. (a) The resist mask for ion-implantation. The cross-patterns are alignment keys for e-beam lithography processes. (b) Silicon nanowires after e-beam lithography process and ICP etching. Each chip consists eight 3μm-long silicon nanowires with widths vary from 70 to 800 nm

employed and the implantation depth in to the top Si-layer was estimated to be about 20 nm, and the concentration of dopant was controlled to about $2 \times 10^{14}/cm^2$. Subsequently, the nanowires (shown in Fig. 18.1b) were defined by electron beam lithography, and formed by ICP etching process. To ensure a good isolation between the devices and the molecular solutions that are to be introduced during the sensing measurements, both Si nanowires and Si leaders were covered by a $\sim$10 nm-thick thermally grown silicondioxide, which was formed in 900 °C dry oxygen ambient for 20 min. The SiO_2 layer could not only provide electrical isolation with solution, but also act as a buffer layer for later surface chemical modification. As the last step of the device fabrication, the surface oxide layer on the outside contact pads was removed by buffer oxide etch process, and 50 nm-thick Au buffered by 10 nm-thick Ni serving as Ohmic contact metals were thermally deposited using the same resist mask.

18.1.2 Characterizations of Unmodified Silicon Nanowires

The fabricated silicon nanowires were characterized by DC electric measurement, and showed field effect transistor behaviors. A bias voltage V_d was applied on the drain electrode while the source electrode was grounded, and a gate voltage V_g was applied on the back-gate electrode fabricated on the backside of the bulk Si chip. Shown in Fig. 18.2a is the measured silicon nanowire, which is 200 nm wide and 3μm-long. Figure 18.2b shows drain-source current I_d as a function of back-gate voltage V_g measured at various drain voltage V_d. The Current I_d increased with decreasing gate-voltage V_g, indicating that the silicon nanowire is of p-type. Figure 18.2c shows a comparison between silicon nanowires of different widths. It is found that the drain currents at a given biasing condition were not proportioned to the wire width; this may be attributed to the inevitable presence

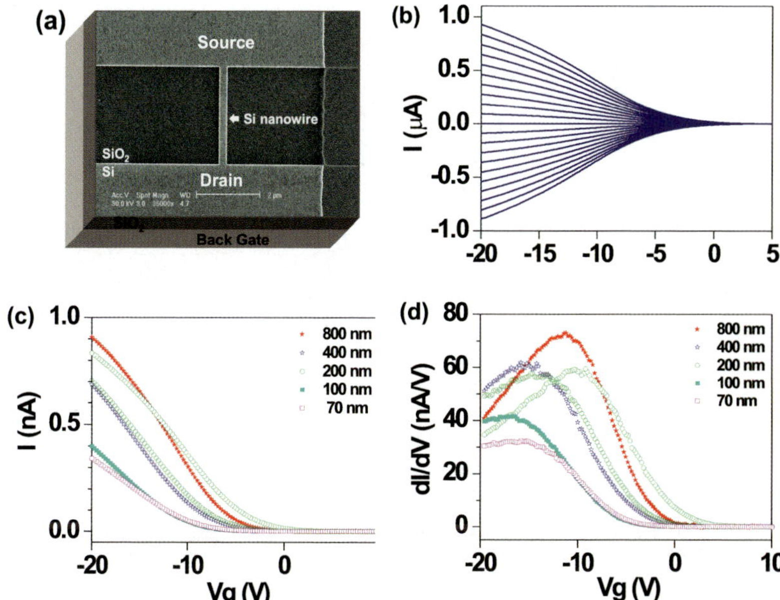

Fig. 18.2 (a) Schematic of the device structure with an SEM image of the measured silicon nanowire which is 200 nm in width and 3 µm in length. (b) The $I_d - V_g$ characteristics measured with the drain voltage Vd ramped from -1 to 1 V with a step of 0.1 V. The device shows p-type FET behavior. (c) $I_d - V_g$ characteristics for wires with different widths (70, 100, 200, 400, 800 nm), represented by different colors. (d) Gate voltage dependence of the transconductance. The charge sensitive is at the maximum at the gate voltage corresponding to the maximum transconductance

of random distribution of impurities/defects. Figure 18.3d shows derivative of the curves in Fig. 18.3c in respect to the gate voltage, i.e. dI_d/dV_g, which is named as transconductance. The carrier mobility is related to the transconductance as $\mu = (L^2/V_dC)(dI_d/dV_g)$, where L is the wire length and C is the wire capacitance (Vashaee et al., 2006) to the backgate electrode. This equation, together with estimated capacitance values and the data shown in Fig. 18.3d, yields a mobility value of about $40\,\text{cm}^2/\text{V s}$. This is small compared to the bulk silicon values, but is comparable with that of SOI-based silicon films (Ohata et al., 2006).

18.1.3 Aptes Modification Procedures

In this process, chips with Si-nanowires were first oxidized with 2% cholic acid ($C_{24}H_{40}O_5$, Fluka, BioChemika, purity $\geq$ 99%) in alcohol for 12 h to generate $-OH$ on the SiO_2 surface. As shown in Fig. 18.3, the chips were then silanized with 2% APTES (from Sigma-Aldrich, purity 99%) in acetone at $80\,°C$ for 2 h to form a single layer of APTES with the amino functional group ($-NH_2$) on the surface (Fixe et al., 2006; Sakata and Miyahara, 2005; Xiang et al., 2004). The devices were subsequently cleaned with distilled H_2O, blow-dry with nitrogen to remove

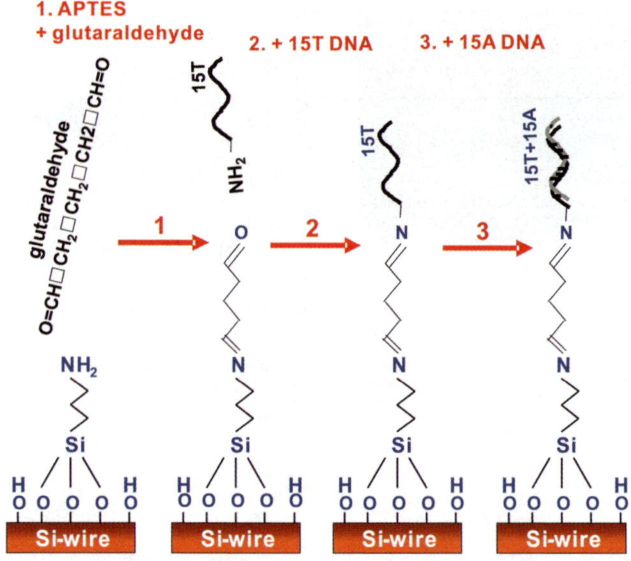

Fig. 18.3 A schematic drawing of the APTES surface modulation procedures

unbound APTES and dried at 110 °C for 1 h. Self-assembly of APTES layer was confirmed with a fluorescent microscope using fluorescein-5-isothiocyanate (FITC) as dye which reacts specifically with NH_2 to form covalent bonds.

18.1.4 Immobilization of Capture DNA Molecules

In this process, 15-mer poly-T ssDNA molecules acting as the capture DNA for subsequent hybridization test was attached to the surface of a APTES-modified 70 nm-wide *n*-type silicon nanowire. The amino functionalized substrates were modified with cross-linker glutaraldehyde (25% in 0.1M phosphate buffer, pH 7.4) at room temperature for 1 h to let one terminal of glutaraldehyde forming an amido bonds -CONH- with APTES and left the other terminal available for further immobilization with single strand DNA (ss-DNA). After cross-linking, 1 μM 15-mer poly-T ssDNA probe solution was added and stayed for 2 h to allow for linkage between DNA probe at the 5′ end and the terminal aldehyde group on the wire surface. The unbound DNA molecules were then washed away by large amount of phosphate buffered solution (PBS).

18.1.5 Fabrication of Microfluidic Channels

In order to detect molecule reactions, such as DNA hybridization, the silicon nanowire sensors were placed inside a microfluidic channel. The channels were made of polydimethylsiloxane (PDMS) using SU8 thick photoresist as mold

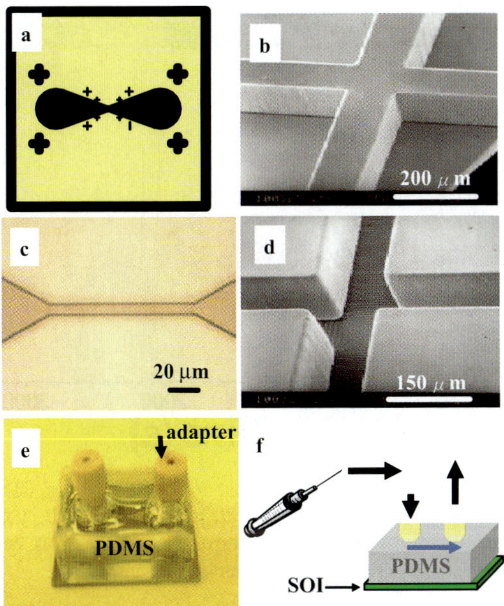

Fig. 18.4 (a) Photo-mask of the microfluidic channel. (b) SEM image and (c) Optic microscope image of the SU8 mold. (d) SEM image of the finished PDMS structure which is a reverse of the mold shown in (b). The images (b) and (d) are taken at the cross-marks in (a). (e) PDMS slab with microfluidic channel on the bottom and tube adapter on the top. (f) Schematic of the fluid system setup

(Ryu, 2004). The fabrication procedures are as follows: The SU8 mold was made on silicon chips by standard photolithography process. SU8-2007 resist was spun on silicon chip at a speed of 3000 rpm for 30 s, which gave a thickness of about 8 μm after a 30 min curing at 65 °C, followed by a second 4-h curing at 95 °C. After exposure with the photo-mask shown in Fig. 18.4a, and post-exposure baked at 95 °C for 30 min, the resist was subjected to a develop process. This produced SU8 molds shown in Fig. 18.4(b) and 4c. Separately, PDMS base was mixed with curing agent with a ratio of 10:1, and was poured into a stainless steel container with the SU8 mold placed at the bottom. The resultant PDMS fluidic channel is shown in Fig. 18.4(d), which was 8 μm high and 10 μm wide. As shown in Fig. 18.4e, the PDMS slab was incorporated with two 800 μm-OD Teflon fluid tube adapts on the top, which connected the tube to the microfluidic channels. Figure 18.4f is a schematic drawing of the fluid circuit.

18.1.6 Detection of DNA Hybridization

The 70 nm-wide silicon nanowire with capture ssDNA on the surface was used as a sensor for hybridization of complementary ssDNA. Real-time resistance response to the DNA hybridization is shown in Fig. 18.5. The resistance value was taken using a lock-in amplifier with an AC excitation of 80 mV and 0.777 Hz. Initially, the silicon

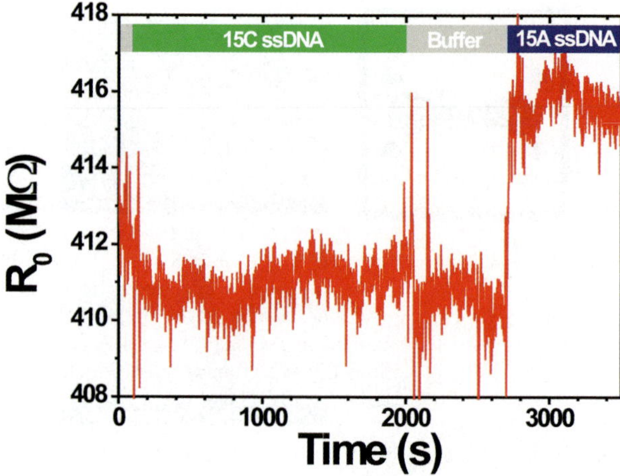

Fig. 18.5 Measured nanowire resistance in response to the injection of ssDNA molecules. Prior to the measurement the wire was immobilized with poly-T ssDNA. Poly-C DNA (20 pM) and buffer solution were added at $t = 100$ and 2000 s, respectively, as controls. At 2700 s ploy-A DNA (20 pM) was added, and an abrupt increase in resistance was observed

nanowire device was immersed in a buffer solution of pH 8. At $t = 100$ s, 20 pM 15-mer poly-C ssDNA molecules was introduced, which should not interact with the capture poly-T ssDNA, and the wire showed no obvious change in the resistance as expected. The unpaired poly-C ssDNA was then removed by buffer solution at $t = 2000$ s. Then the target DNA, that is 15-mer poly-A ssDNA molecules with a concentration of 20 pM, was introduced at $t = 2700$ s and the wire showed a rapid increase in the resistance. Since DNA molecules are negatively charged in pH 8 buffer solution, hybridization of two ssDNA simply implies an increased negative charge on the wire surface, which then results in depletion of conduction electrons in the n-type silicon nanowires. This is manifested in an increased wire resistance as shown in our experiment. This result demonstrates detection of hybridization between very short DNA molecules using silicon nanowires.

18.2 Conclusions

In this study, we developed techniques for fabrication of nanometer-wide SOI-based silicon wires and PDMS micro fluid channels. Furthermore, conditions for APTES modification of the silicon nanowire surface as well as immobilization of capture DNA molecules were explored. With these technologies, we investigated the potential of using silicon nanowires as a detector for DNA hybridization. Our experiment result showed that detection of hybridization between complementary ssDNA with lengths as short as 15 mer is possible with a 70nm-wide SOI-based silicon nanowire.

References

Fixe, F., Dufva, M., Telleman, P. and Christensen, C. B. V., 2004, Functionalization of poly(methyl methacrylate) (PMMA) as a substrate for DNA microarrays. Nucleic Acids Research, **32,** p. e9.

Li, Z., Chen, Y., Li, X., Kamins, T. I., Nauka, K. and Williams, R. S., 2004, Sequence-Specific Label-Free DNA Sensors Based on Silicon Nanowires. Nano Letters, **4,** pp. 245–246.

Ohata, A., Cristoloveanu, S. and Cassé, M., 2006, Mobility comparison between front and back channels in ultrathin silicon-on-insulator metal-oxide- semiconductor field-effect transistors by the front-gate split capacitance-voltage method. Applied Physics Letters, **89,** pp. 032104–032106.

Ryu, K. S., Wang, X., Shaikh, K. and Liu, C., 2004, A Method for Precision Patterning of Silicone Elastomer and Its Applications. Journal of Microelectromechanical Systems, 13, pp. 568–575.

Sakata, T. and Miyahara, Y., 2005, Potentiometric Detection of Single Nucleotide Polymorphism by Using a Genetic Field-effect transistor. ChemBioChem, **6,** pp. 703–710.

Vashaee, D., Shakouri, A., Goldberger, J., Kuykendall, T., Pauzauskie, P. and Yang, P., 2006, Electrostatics of nanowire transistors with triangular cross sections. Journal of Applied Physics, **99,** pp. 054310–054314.

Xiang, J., Zhu, P., Masuda, Y. and Koumoto, K., 2004, Fabrication of Self-Assembled Monolayers (SAMs) and Inorganic Micropattern on Flexible Polymer Substrate. Langmuir, **20,** pp. 3278–3283.

Chapter 19
From Simple Molecules to Molecular Functional Materials and Nanoscience

Vivian Wing-Wah Yam and Keith Man-Chung Wong

19.1 Introduction

The rational design and synthesis of molecular and nano-scale materials have attracted growing attention in recent years owing to their enormous and unpredictable potentials in molecular devices and machineries, and nanoscopic sciences. Apart from the design of the discrete molecules and an understanding of the structure-property relationships based on the molecules themselves, an exploration of the work beyond the molecular level towards the supramolecular and nanoscopic scale would represent challenging areas of research.

By combining the advantages of high synthetic versatility and easy processibility in the molecular properties, together with an understanding of the structure-property relationship established, molecular functional materials could be readily prepared by the approach of "function by design". Moreover, it would also be interesting to study how the properties in the bulk might be related to those at the nano and molecular level and whether some insights could be provided through a structure-property relationship study. All these would be beneficial to the future design of functional materials in the area of nanoscience and materials.

In this report, the design and synthesis of a number of transition metal complex molecules including Cu(I), Ag(I), Au(I), Au(III), Re(I), Pt(II) and Ru(II) have been described. All the complexes have been found to exhibit strong luminescence. Their fundamental photophysical properties including electronic absorption and luminescence behaviors have been studied and their spectroscopic origins have been elucidated. Through rational design, judicious functionalization and assembly strategies, some of these metal complexes could find potential applications as various molecular functional materials, and as templates for the preparation of nano-sized materials.

Centre for Carbon-Rich Molecular and Nano-Scale Metal-Based Materials Research, and Department of Chemistry, The University of Hong Kong, Pokfulam Road, Hong Kong, P. R. China

Z. Tang, P. Sheng (eds), *Nano Science and Technology.*
© Springer 2008

19.2 Transition Metal Chalcogenides and Pnictogenide

19.2.1 Tetranuclear Copper(I) and Silver(I) Chalcogenides and Pnictogenide

Despite a large number of transition metal chalcogenide cluster and organometallic aggregates of d^{10} metal centers containing short metal-metal interactions are known (Grohmann et al., 1995), most of them are insoluble and the photophysical and photochemical studies of the related species are relatively less explored. We have successfully synthesized a series of tetranuclear copper(I) and silver(I) complexes with μ_4-bridging chalcogenides, $[M_4(P^\wedge P)_4(\mu_4\text{-}E)]^{2+}$ {M = Cu, E = S, Se; M = Ag, E = S, Se, Te; $P^\wedge P$ = bis(diphenylphosphino)methane (dppm), bis[bis(4-methylphenyl)-phosphino]methane (dtpm)}, from the reaction of $[M_2(P^\wedge P)_2(MeCN)_2]^{2+}$ with Na_2E or Li_2E (Scheme 19.1) (Yam et al., 1993, 1996b,c, 2001b). All the complexes have been structurally characterized by X-ray crystallography, with the four metal centers, in each case, arranged in a distorted rectangular geometry bridged by a μ_4-chalcogenide ligand. Short $Cu \cdots Cu$ and $Ag \cdots Ag$ contacts in the range of 2.869–3.271 and 3.038–3.357 Å, respectively, are observed. A related isostuctural tetranuclear copper(I) pnictogenide complex, $[Cu_4(dppm)_4(\mu_4\text{-PPh})]^{2+}$, has also been prepared by reacting $PhPH_2$ with $[Cu_2(dppm)_2(MeCN)_2]^{2+}$ in the presence of Na(acac) (Deveson et al., 1997). These complexes were found to exhibit intense long-lived luminescence at 527–718 nm with excitation wavelength at $\geq$ 350 nm in the solid state. The transition associated with this luminescence is assigned to originate from the ligand-to-metal charge transfer (LMCT) $[E^{\tilde{2}}/PPh^{\tilde{2}} \rightarrow M_4]$ triplet excited state, with some mixing of a

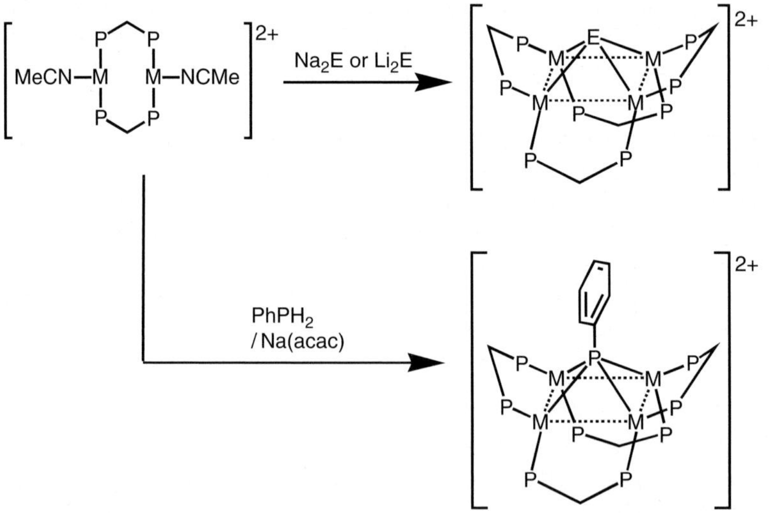

Scheme 19.1 Synthetic scheme for tetranuclear copper(I) and silver(I). chalcogenide and phosphinidene complexes

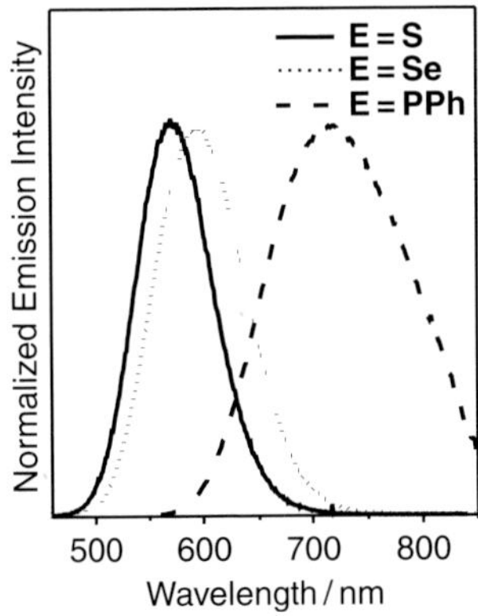

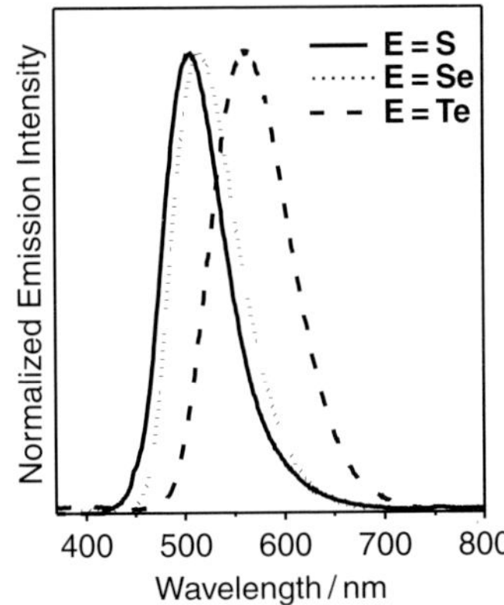

Fig. 19.1 Solid-state emission spectra of tetranuclear copper(I) and silver(I) chalcogenide and phosphinidene complexes

metal-centered (MC) state. In general, the solid-state emission energies were found to follow the order: $E = S > Se > Te > PPh$ and $M = Ag > Cu$ (Fig. 19.1). This energy trend is in line with the ionization energies of the chalcogens or pnictogen and supports an excited-state assignment of large LMCT $[E^{\tilde{2}}/PPh^{\tilde{2}} \rightarrow M_4]$ character. Depending on the metal centers and the chalcogens or pcictogen in these metal clusters, different luminescence colors were observed (Table 19.1).

19.2.2 High Nuclearity Gold(I) Sulfido Clusters

Despite the fact that numerous copper(I) and silver(I) chalcogenide clusters (Canales et al., 1996; Corrigan and Fenske, 1997; Yam and Lo, 1997) are known, examples of chalcogenide clusters of gold(I), which belongs to the same group as copper(I) and silver(I), are limited (Yam et al., 1999). A series of high-nuclearity gold(I) sulfido complexes with bridging diphosphine ligands have been synthesized and isolated by us, with the general formulae of $[Au_{12}(\mu\text{-}P^{\wedge}P)_6(\mu_3\text{-}S)_4]^{4+}$, $[Au_{10}(\mu\text{-}PNP)_4(\mu_3\text{-}S)_4]^{2+}$ and $[Au_6(\mu\text{-}PNP)_3(\mu_3\text{-}S)_2]^{2+}$ $[P^{\wedge}P = \text{bis(diphenylphosphino)-methane (dppm)}; PNP = \text{bis(diphenylphosphanyl)-}n\text{-propylamine } (Ph_2PN(^nPr)PPh_2)]$ (Yam et al., 1996a, 2000a, 2001a). Depend-

Table 19.1 Emission color of tetranuclear copper(I) and silver(I) chalcogenide and pnictogenide complexes

M	E	Luminescence color
Cu	S	Orange-yellow
Cu	Se	orange
Cu	PPh	Red
Ag	S	Green
Ag	Se	greenish-yellow
Ag	Te	orange-yellow

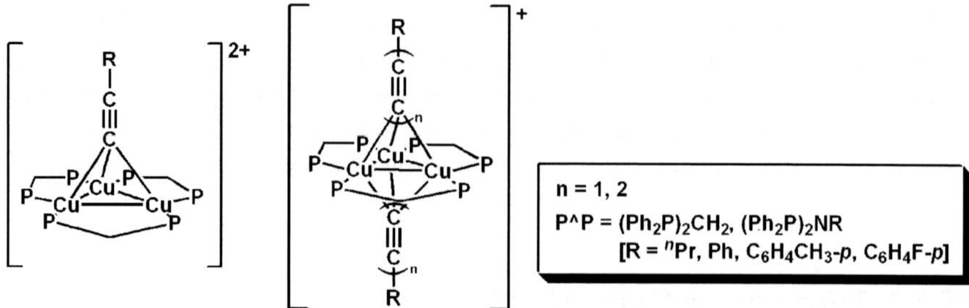

Scheme 19.4 Molecular structures of $[Cu_3(\mu - P^\wedge P)_3\{\mu_3 = \eta^1\text{-}(C \equiv C)_n R\}]^{2+}$ and $[Cu_3(\mu - P^\wedge P)_3\{\mu_3 = \eta^1\text{-}(C \equiv C)_n R\}_2]^+$

^{n}BuLi in the appropriate ratio under anaerobic and anhydrous conditions ($P^\wedge P = Ph_2PCH_2Ph_2$ (dppm); $Ph_2PN(R')PPh_2$, $R' = Ph$, $^n Pr$, $C_6H_4\text{-}CH_3\text{-}p$) (Scheme 19.4) (Lo et al., 2004; Yam et al., 1993b, 1997). The interesting photophysical and photochemical properties of a number of luminescent polynuclear copper(I) alkynyl complexes have recently been studied. In all cases, the excited states of the complexes have been proposed to bear a substantial ^{3}LMCT [$(C \equiv C)_n R \rightarrow Cu$] character, and probably mixed with a metal-centered d-s triplet-state. The emission energies were found to be dependent on the R substituent group with same alkynyl chain length and lower emission energies were observed for the complexes with the stronger electron-donating substituents on the alkynyl ligand. For instance, the emission energies follow the trend: $[Cu_3(\mu\text{-}dppm)_3\{\mu_3 - \eta^1\text{-}(C \equiv C)_2 C_6 H_4\text{-}OCH_3\text{-}p\}_2]^+$ < $[Cu_3(\mu\text{-}dppm)_3\{\mu_3 - \eta^1\text{-}(C \equiv C)_2 C_6 H_4\text{-}CH_3\text{-}p\}_2]^+$ < $[Cu_3(\mu\text{-}dppm)_3\{\mu_3 - \eta^1\text{-}(C \equiv C)_2 C_6 H_5\}_2]^+$ (Fig. 19.2), which is in line with the assignment of ^{3}LMCT[$(C \equiv C)_2 R \rightarrow Cu$] character. Moreover, their emission energies could also be tuned by the variation of the alkynyl chain length. The greater is the number of alkynyl units that the complex

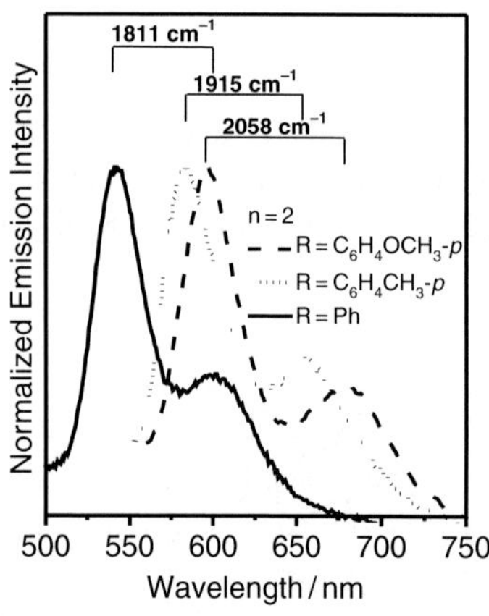

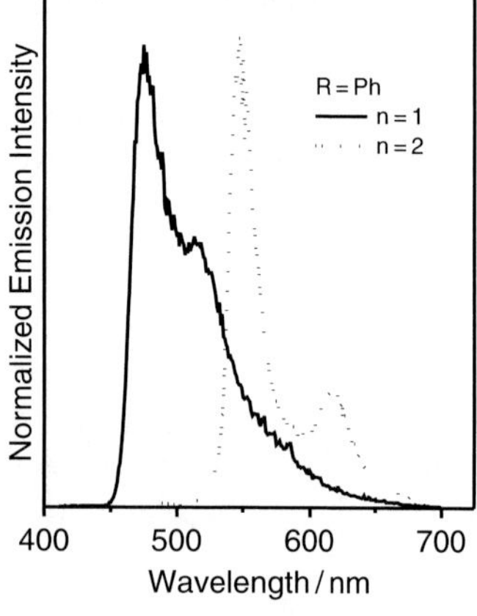

Fig. 19.2 Normalized solid-state emission spectra of $[Cu_3(\mu\text{-}dppm)_3\{\mu_3 = \eta^1\text{-}(C \equiv C)_n R\}_2]^+$

possesses, the lower is the emission energy observed (Fig. 19.2). This energy trend is also supportive of the assignment of a mixed metal cluster-centered d–s/LMCT state for the emission origin since the π-electron donating ability decreases on going from the diynyl moiety to the monoynyl counterparts.

19.3.3 Rhenium(I) Diimine Alkynyls

The photoluminescence properties of tricarbonylrhenium(I) diimine complexes have also been extensively studied and a metal-to-ligand charge-transfer (MLCT) $[d\pi \rightarrow \pi^*(\text{diimine})]$ excited state were assigned for their emission origin (Wrighton and Morse, 1974; Fredericks et al., 1979). In view of the well documented MLCT excited state chemistry of rhenium(I) diimine complexes, a project of the synthesis and photophysical studies of the related luminescent alkynylrhenium (I) diimine complexes was launched.

Luminescent mono- and dinuclear rhenium(I) alkynyl complexes, $[\text{Re(CO)}_3(N^\wedge N)-(C\equiv C)_n R]]$ and $[(N^\wedge N)(CO)_3\text{Re}-(C\equiv C - C \equiv C)_n-\text{Re(CO)}_3(N^\wedge N)]$ ($n = 1$; N–N = bpy, $^t\text{Bu}_2$bpy, Me$_2$bpy, phen) were first synthesized in our laboratory (Scheme 19.5) (Yam et al., 1995, 1996d, 1998, 2000b; Yam, 2001). Intense orange phosphorescence was observed upon excitation at $\lambda \geq 350$ nm in the solid state and in fluid solution of these rhenium(I) alkynyls. Such luminescence is ascribed to be originated from a ^{3}MLCT $[d\pi(\text{Re}) \rightarrow \pi^*(N^\wedge N)]$ excited state, with some mixing of a $[\pi(C\equiv C) \rightarrow \pi^*(N^\wedge N)]$ ligand-to-ligand charge transfer (LLCT) character. The emission energies have been found to be dependent on the nature of both the diimine and alkynyl ligands. Higher emission energy was observed for the complexes containing better electron-donating substituents on the diimine ligand for the complexes with the same alkynyl ligand (Yam, 2001), which is in agreement with the ^{3}MLCT assignment. Upon extending the carbon chain in these complexes, the emission energy was found to shift to the blue (Yam et al., 2000b) and this is an unusual phenomenon in contrast to the common concept of tuning the emission energy to the red by extending the number of acetylenic units in organic polyynes.

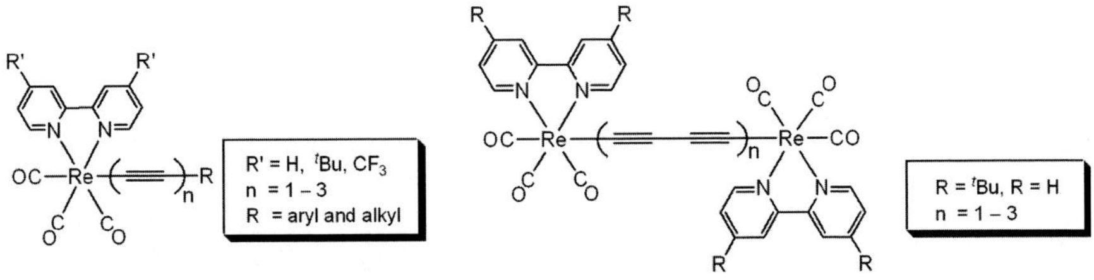

Scheme 19.5 Molecular structures of luminescent mono- and dinuclear rhenium(I) alkynyl complexes

19.3.4 Platinum(II) Terpyridyl Alkynyls

Platinum(II) polypyridyl complexes have been widely studied for their luminescence behavior due to their intriguing photophysical and spectroscopic properties resulting from their square planar coordination geometry, which permits easy access to important frontier orbitals of the metal center, $d^8 - d^8$ metal–metal interactions as well as $\pi - \pi$ interactions (Miskowski and Houlding, 1989; Chan et al., 1994; Marsh et al., 1995; McMillin and Moore, 2002). As a continuation of our great interests in luminescent transition metal alkynyls, a series of luminescent alkynylplatinum(II) terpyridyl complexes, $[Pt(N^\wedge N^\wedge N)(C \equiv C-R)]^+$ ($N^\wedge N^\wedge N$ =terpyridyl ligand; R = alkyl, aryl or C≡CH) (Yam et al., 2001c, 2002b), have been synthesized from the substitution reaction of the labile acetonitrile group on the precursor complex $[Pt(N^\wedge N^\wedge N)(MeCN)]^{2+}$ by the alkynyl group (Scheme 19.6).

The platinum(II) terpyridyl complexes with various substituted phenylethynyl groups, $[Pt(trpy)(C{\equiv}CC_6H_4\text{-}R\text{-}p)]^+$ (R = H, Cl, NO_2, CH_3, OCH_3, NH_2, NMe_2, $N(CH_2CH_2OCH_3)_2$) (Wong et al., 2004), in acetonitrile solution absorb strongly at 286–350 nm and less intensely at 412–546 nm. The low-energy absorption bands are assigned to a $[d\pi(Pt) \rightarrow \pi^*(N^\wedge N^\wedge N)]$ metal-to-ligand charge-transfer (MLCT) transition, with mixing of some $[\pi(C \equiv C) \rightarrow \pi^*(N^\wedge N^\wedge N)]$ ligand-to-ligand charge-transfer (LLCT) contribution. In general, the stronger the electron-donating ability of the alkynyl ligand is, the lower is the energy of the lowest energy absorption. Therefore, such absorption energy follows the trend with $R = NO_2 > Cl > H > CH_3 > OCH_3 > NH_2 > N(CH_3)_2 > N(CH_2CH_2OCH_3)_2$, which is in line with the assignment of a transition of MLCT/LLCT admixture. Unlike the chloro-counterpart, $[Pt(trpy)Cl]^+$, which is non-emissive in the solution state at ambient temperature, most of the platinum(II) terpyridyl alkynyl complexes were found to exhibit luminescence at 560–665 nm in acetonitrile solution at 298 K. An origin of a $[d\pi(Pt) \rightarrow \pi^*(N^\wedge N^\wedge N)]^3$MLCT excited state, with some mixing of a $[\pi(C{\equiv}CR) \rightarrow \pi^*(N^\wedge N^\wedge N)]^3$LLCT state, is assigned to this emission. Similar to the electronic absorption studies, the emission energies of the complexes in solution were found to depend on the nature of the substituents on the phenyl ring of the alkynyl ligands, i.e. the stronger the electron-donating ability of the alkynyl ligand is, the lower is the emission energy. However, for the complexes with stronger electron-donating substituents, such as methoxy and amino groups, on the phenyl ring of the alkynyl ligand, no emission was observed in acetonitrile solution at 298 K. The emission is believed to be quenched by the methoxy or amino group via

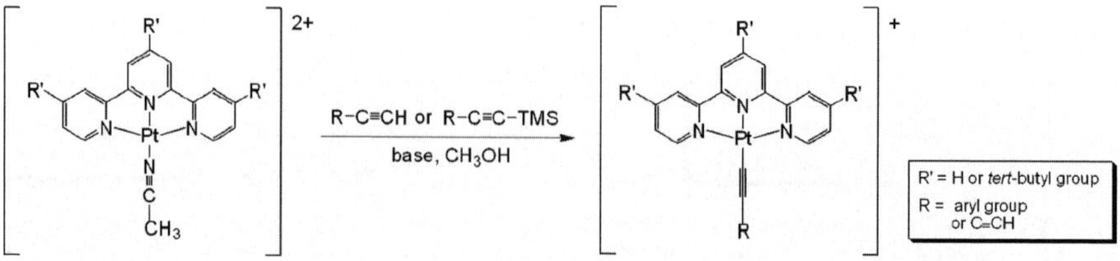

Scheme 19.6 Synthetic scheme for luminescent alkynylplatinum(II) terpyridyl complexes

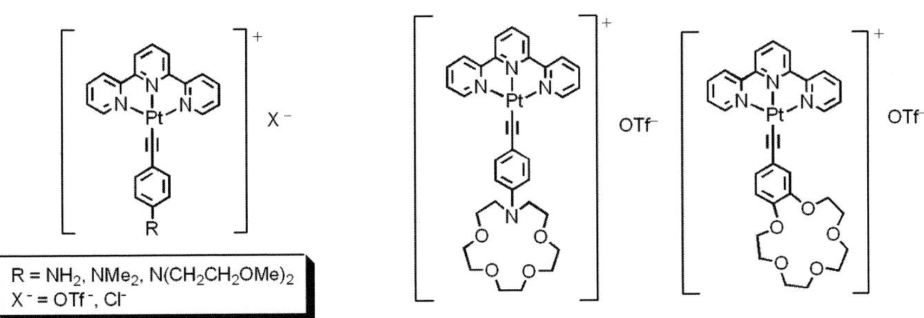

Scheme 19.7 Molecular structures of platinum(II) terpyridyl alkynyl complexes with basic amino functionalities and with crown ether pendants

photo-induced electron transfer or by the presence of the low-lying non-emissive ^{3}LLCT state.

In view of the sensitive dependence of the absorption and emission energy on the R substituents of the alkynyl ligands, functionalization of the alkynyl substituents could afford the generation of various molecular functional materials. A series of platinum(II) terpyridyl alkynyl complexes that have been derivatized with basic amino functionalities, $[Pt(trpy)(C\equiv C\text{-}C_6H_4\text{-}NR_2\text{-}p)]X$ (X = OTf$^-$, or Cl$^-$; R = H, CH$_3$, CH$_2$CH$_2$OCH$_3$), and with benzo-15-crown-5 and N-phenylaza-15-crown-5 pendants, $[Pt(trpy)(C\equiv C\text{-benzo-15-crown-5})]OTf$ and $[Pt(trpy)(C\equiv C\text{-}N\text{-phenylaza-15-crown-5})]OTf$, have been synthesized and characterized (Scheme 19.7) (Wong et al., 2005a; Tang et al., 2005).

Their photophysical responses at various acid or metal cation concentrations were studied. The complexes with basic amino functionalities exhibit dramatic and reversible color change and luminescence switching upon addition of acid (Fig. 19.3), as a result of the protonation of the amino group which decreases the electron-donating ability of the alkynyl ligand and results in a blue shift of the

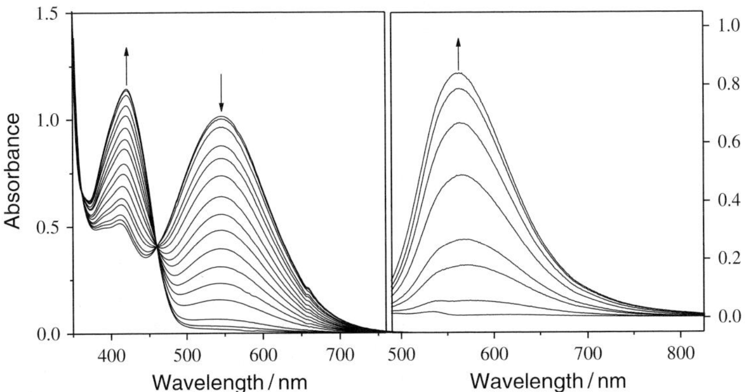

Fig. 19.3 UV-vis absorption (left) and emission (right) spectral changes of [Pt(trpy) $(C \equiv CC_6H_4\text{-}N(CH_2CH_2OCH_3)_2\text{-}p)]$(OTf) in MeCN (0.05 M ^{n}Bu₄NPF₆) with increasing p-toluenesulfonic acid content

low-energy LLCT absorption and the corresponding shift of the ^{3}LLCT state to higher energy. The reversibility of color changes was studied upon alternate addition of p-toluenesulfonic acid and triethylamine. On the other hand, the complexes with crown ether pendants exhibit electronic absorption spectral changes upon addition of metal ions, and in some cases these are accompanied by visual color changes. The observed blue shift in the low-energy MLCT absorption band upon ion-binding is attributed to result from the complexation of metal ions to the crown ether moiety, which would decrease the electron-donating ability of the alkynyl ligand, leading to the lowering of the dπ(Pt) orbital energy. These findings suggest that with a suitable design of the ligands and complexes and a sound understanding of the spectroscopic properties of these complexes, promising potential colorimetric and luminescence pH and metal cation sensors could be developed.

Two luminescent alkynylplatinum(II) terpyridyl complexes, [Pt(^{t}Bu$_3$trpy)(C≡C-C$_6$H$_4$-X − p)](OTf)(X = NCS and NHCOCH$_2$I), have also been designed and synthesized (Scheme 19.8) (Wong et al., 2004). Their photophysical properties have been studied and both exhibited strong luminescence in various media. Similar to the other platinum(II) terpyridyl alkynyl complexes, the emission origin has been tentatively assigned as a dπ(Pt) → π^*(^{t}Bu$_3$trpy)^{3}MLCT excited state, with some mixing of a π(C≡CR) → π^*(^{t}Bu$_3$trpy)^{3}LLCT character. Human serum albumin (HSA) has been labeled by the ready reaction of the isothiocyanate and iodoacetamide functional groups in these two complexes to afford the respective bioconjugates, through the formation of thiourea and thioether linkage. Both bioconjugates in 50 mM Tris-Cl buffer (pH 7.4) are highly colored and exhibit luminescence in the visible region upon photoexcitation. The bioconjugate resulting from the isothiocyanate complex is found to emit with a different luminescence color when compared to its parent label (Fig. 19.4).

Interesting and rich polymorphic behavior was also observed in the platinum(II) terpyridine butadiynyl complex, [Pt(trpy)(C≡CC≡CH)]OTf, which was found to exist in two forms, i.e. a dark green form and a red form (Yam et al., 2002b). Both of which have been structurally characterized and shown to exhibit different crystal packing arrangements. The dark green form exists as a linear-chain with

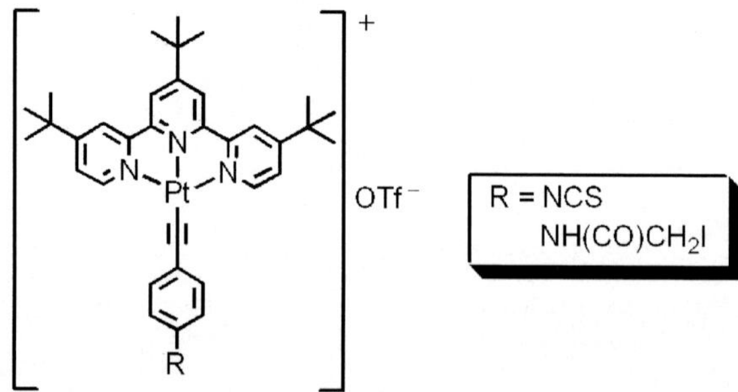

Scheme 19.8 Molecular structures of platinum(II) terpyridyl alkynyl complexes for biolabeling studies. Reprodued with permission from (32)

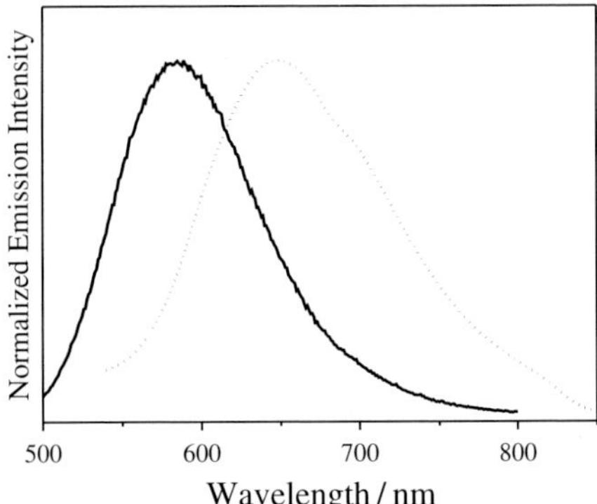

Fig. 19.4 Emission spectra of [Pt(^{t}Bu$_3$trpy)(C $\equiv$ C-C$_6$H$_4$-NCS-p)](OTf) (—) in acetonitrile solution and its HSA bioconjugate ($\cdots\cdots$) in 50 mM Tris-Cl pH 7.4 buffer solution

the platinum atoms equally spaced, with short intermolecular Pt$\cdots$Pt contacts of 3.388 Å. The red form, on the other hand, shows dimeric structure with alternating Pt$\cdots$Pt distances of 3.394 and 3.648 Å. Dissolution of both forms in acetone or acetonitrile gave the same absorption spectrum, indicating that the same molecular species exists in the solution state. There was a dramatic color change with a tremendous emission enhancement in solution upon varying the composition of the solvent. Upon increasing diethyl ether content in solution, the color of the solution changed dramatically from yellow to green to blue, with a new absorption band formed at 615 nm concomitant with a drop in absorbance at 416 nm (Fig. 19.5), and a tremendous emission enhancement at 785 nm in the near-infrared region was observed (Fig. 19.5). Both the emission enhancement and new absorption band formation were ascribed to be derived from a metal-metal-to-ligand charge transfer (MMLCT) origin of aggregated species; the formation of which was facilitated by the propensity of this class of compounds to form Pt$\cdots$Pt and $\pi\cdots\pi$ interactions.

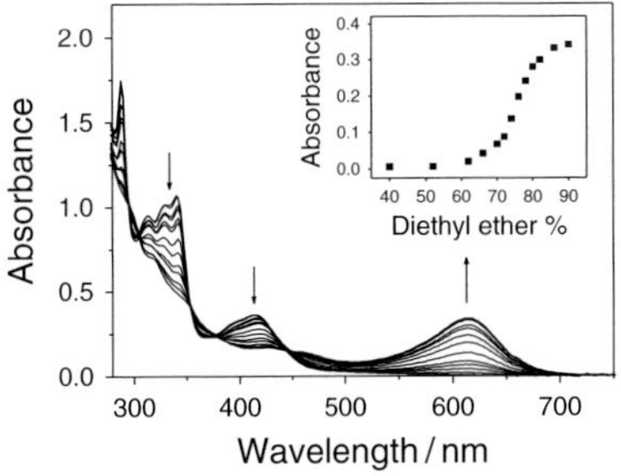
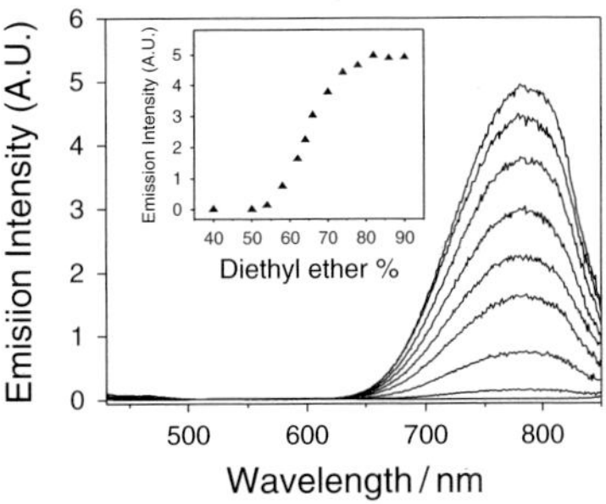

Fig. 19.5 UV-vis absorption (left) and emission (right) changes of [Pt(tpy)(C $\equiv$ CC $\equiv$ CH)]OTf in acetonitrile with increasing diethyl ether composition. Inset (left): Plot of absorbance vs. diethyl ether composition in acetonitrile at 615 nm (■). Inset (right): Plot of corrected emission intensity as a function of diethyl ether composition (▲)

It is interesting to note that such drastic color changes and emission enhancement arising from the solvent-induced aggregation are dependent on the nature of counter anions (Yam et al., 1993a). Upon solvent-induced aggregation, the solution colors change from yellow to blue in [Pt(trpy)(C$\equiv$CC$\equiv$CH)]OTf, to magenta in [Pt(trpy)(C$\equiv$CC$\equiv$CH)]PF$_6$, to pink in [Pt(trpy)(C$\equiv$CC$\equiv$CH)]ClO$_4$, and to orange in [Pt(trpy)(C$\equiv$CC$\equiv$CH)]BF$_4$. Similarly, emission enhancement was observed at different wavelength upon solvent-induced aggregate formation which demonstrates that counter anions play an important role in governing the degree of aggregation and the extent of interactions within these aggregates.

19.3.5 Tetranuclear Gold(I) Alkynylcalix [4] Crown-6 Complexes

Gold(I) alkynyl systems without phosphine ligands are usually polymeric or oligomeric in nature, resulting in intractable material, and molecular complexes of this type are therefore extremely scarce. Alkynylcalixarenes may serve as versatile ligands for the construction of novel luminescent gold(I) alkynyl supramolecular assemblies by making use of their predefined conformations and preorganized geometries. We have recently synthesized and isolated a series of novel luminescent tetranuclear gold(I) alkynylcalix [4] crown-6 assemblies (Scheme 19.9) (Yip et al., 2004). The X-ray crystal structure shows a double-cage structure, in which the two diethynylcalixcrown ligands are σ-coordinated to two gold(I) centers on the two ends, whereas the other two Au centers are each π-coordinated to the two alkynyl units in a η^2, η^2-bonding fashion. All these complexes show intense luminescence upon excitation at $\lambda \geq 370$ nm. Their solid-state emission spectra show low-energy emission bands at about 590$\tilde{}$620 nm at both 77 K and room temperature. Given the presence of short Au$\cdots$Au distances observed in the crystal structure, the low-energy emission is assigned as derived from states of metal-cluster-centered (ds/dp) character that are modified by Au$\cdots$Au interactions with mixing of metal-perturbed intraligand $\tilde{\pi}\pi^*$(C$\equiv$C) states.

19.3.6 Biscyclometalated Gold(III) Alkynyls

Luminescent gold(III) compounds are rare, with very few exceptions that emit at room temperature in solution (Chan et al., 1994; Yam et al., 1993a). One of the

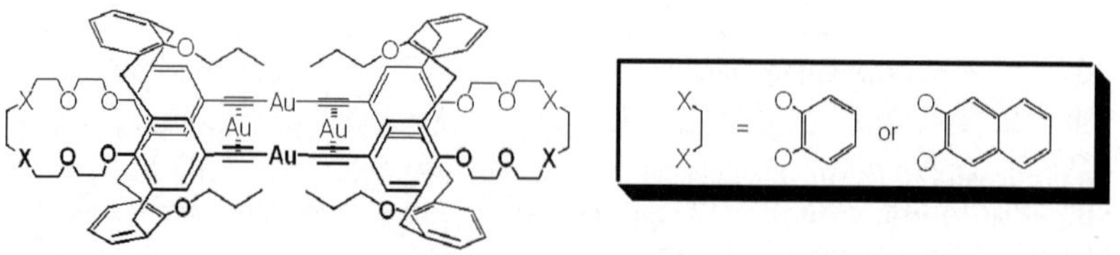

Scheme 19.9 Molecular structures of luminescent gold(I) alkynyl supramolecular assemblies

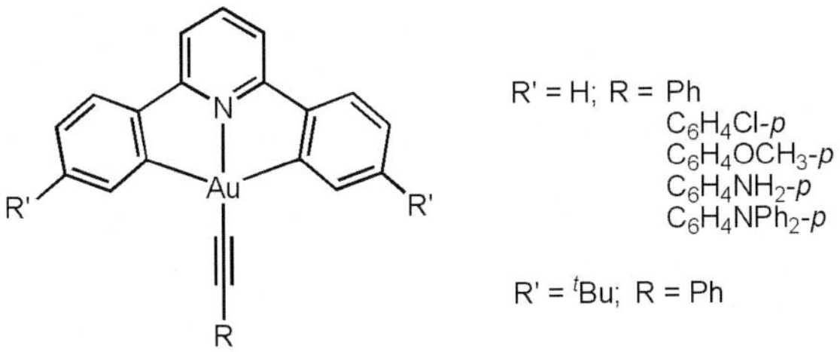

Scheme 19.10 Molecular structures of luminescent alkynylgold(III) complexes. Reproduced with permission from (37)

reasons for the lack of luminescence in gold(III) species is the presence of low-energy $\tilde{d}d$ ligand field (LF) states. As an extension of our great interest in transition metal alkynyl complexes, together with the potential improvement or enhancement of the luminescence behaviors by introduction of strong σ-donating alkynyls to gold(III) center, we recently synthesized a novel class of luminescent alkynylgold(III) complexes, [Au(C^N^C)(C≡CR)] (HC^NCH = 2, 6-diphenylpyridine) (Scheme 19.10) (Yam et al., 2005b). All complexes exhibit an intense vibronic-structured absorption band at $36\tilde{2}412$ nm in dichloromethane. Since the absorption energy was found to be rather insensitive to the nature of the alkynyl ligand, together with the observation of vibrational progressional spacings of $131\tilde{0}1380$ cm$^{\tilde{1}}$ in the low-energy absorption band corresponding to the skeletal vibrational frequency of the C^N^C ligand, the low-energy absorption band is assigned as metal-perturbed intraligand (IL) $\tilde{\pi}\pi^*$ transition of the C^N^C ligand, involving some charge transfer from the phenyl moiety to the pyridyl unit. This assignment is further supported by the observation of this low-energy absorption at lower energy in the complex with two electron-donating *tert*-butyl groups on the phenyl rings of the C^N^C ligand compared with the unsubstituted analogue (Fig. 19.6). Intriguingly, this class of alkynylgold(III) complexes exhibit intense luminescence upon photoexcitation even at ambient temperature. A vibronic-structured emission band with band maximum at around 480 nm was generally observed in dichloromethane solution at room temperature. The luminescence has been assigned as originating from a metal-perturbed IL $^3[\pi-\pi]$ state of the tridentate C^N^C ligand. Similar to the electronic absorption studies, the complex with two*tert*-butyl groups on the C^N^C ligand was found to emit at lower energy than that without such substituents (Fig. 19.6). It is noteworthy that their solid-state emission spectra show a structureless band (550–585 nm) at lower energy than that in solution, which is attributed to dimeric or excimeric emission arising from the $\pi-\pi$ stacking of the C^N^C ligand.

This class of luminescent cyclometalated gold(III) alkynyl complexes has been demonstrated to possess electroluminescence (EL) properties and has been employed as electrophosphorescent emitters or dopants of OLEDs with high brightness and efficiency (Wong et al., 2005b). The color of the EL has been shown to be capable of tuning by a variation in the applied DC voltage as well as the dopant concentration (Fig. 19.7). In one of the devices, the emission color can be tuned from orange to blue through white by applying different DC voltages. On

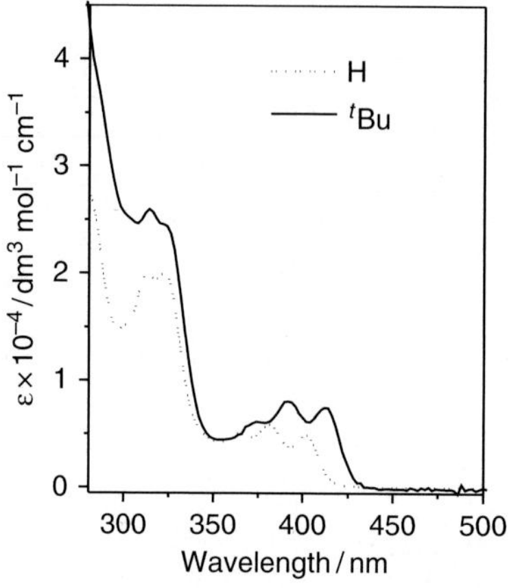
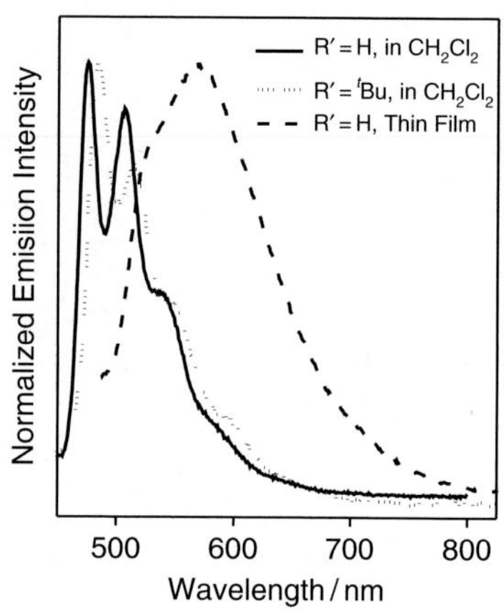

Fig. 19.6 Electronic absorption (left) and emission (right) spectra of [Au(R′C^N^CR′)(CC-Ph)] at 298 K

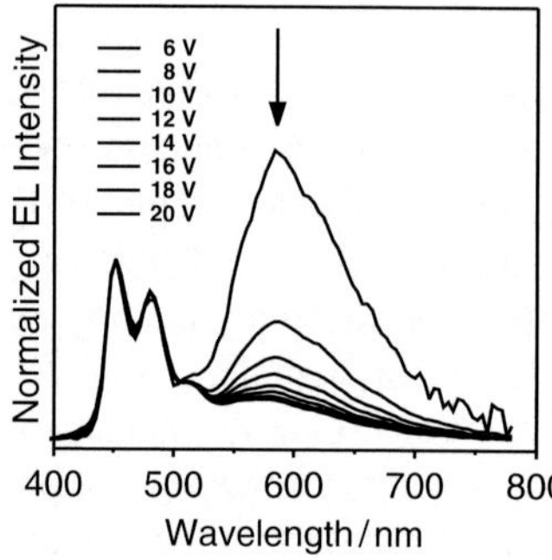
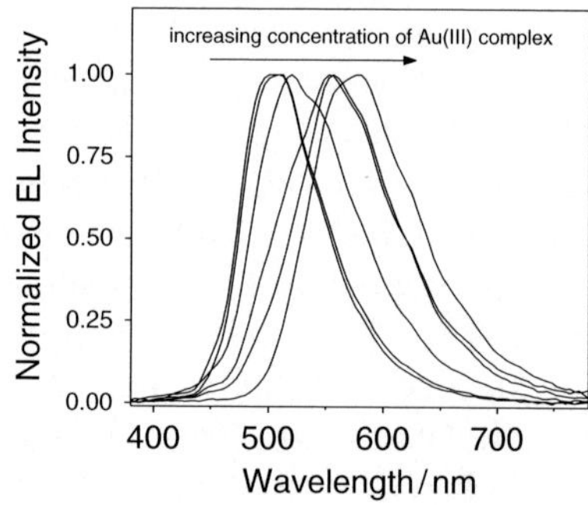

Fig. 19.7 Electroluminescence spectra of OLED devices with Au(III) alkynyl complexes showing the voltage- and dopant concentration-dependent EL properties

the other hand, the EL band maxima are found to red-shift from 500 to 580 nm in another device upon increasing the dopant concentration. A maximum external quantum efficiency of 5.5%, corresponding to a current efficiency of 17.6 cd A^{-1} and luminance power efficiency of 14.5 lm W^{-1}, has been obtained in one of the multilayer OLEDs.

19.4 Template Synthesis of Mesoporous Silicates and Nanoparticles

A series of surfactant ruthenium(II) complexes that possess lyotropic liquid crystalline properties have been synthesized (Scheme 19.11) for the design of new metallomesogens (Yam et al., 2002a). Through a variation of the alkyl chain length and the head group of the metal complex template, metal-containing mesoporous

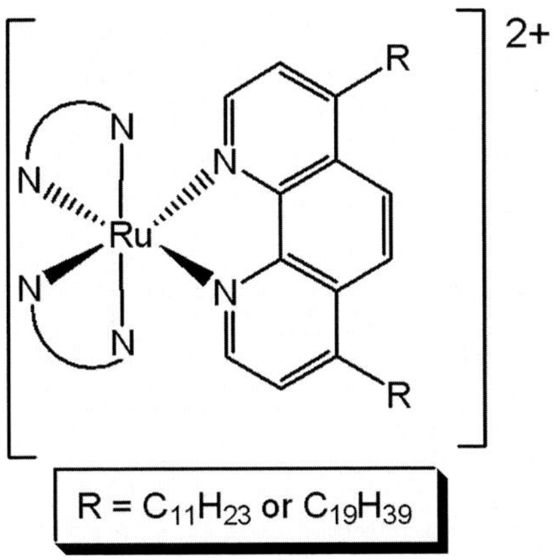

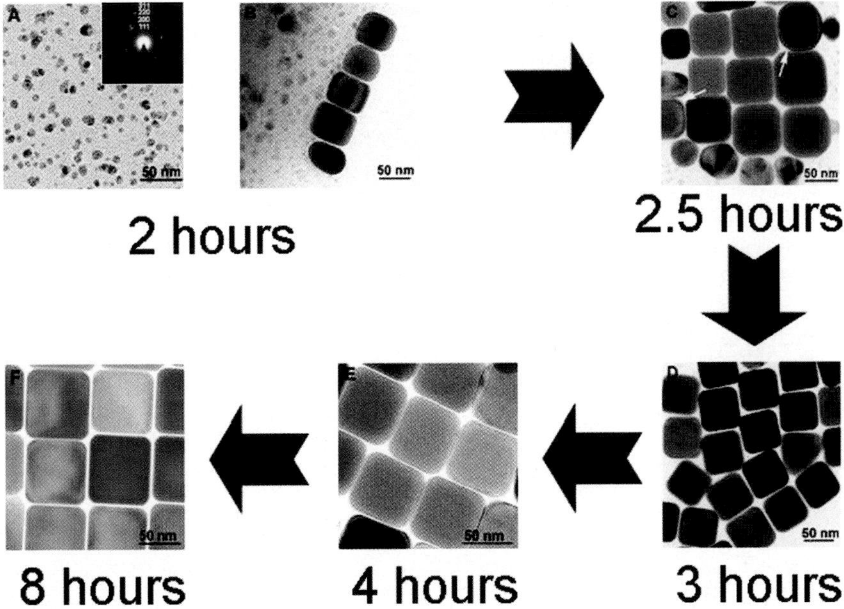

Scheme 19.11 Molecular structures of surfactant ruthenium(II) complexes

silicates with different pore size can be prepared. These materials may have potential applications in catalysis and optical sensor functions.

Recently, we have synthesized monodisperse silver nanocubes with edge length of 55 ± 5 nm, for the first time, on the basis of n-hexadecyltrimethylammonium bromide (HTAB)-modified silver mirror reaction in water (Yu and Yam, 2004, 2005). The addition of HTAB could control the consumption rate of Ag(I), which exists in the form of AgBr. The HTAB is adsorbed on the cube surfaces to generate a

Fig. 19.8 TEM images of Ag NPs at different synthesis time. Reproduced with permission from *(41)*

regular interparticle spacing, which avoids random aggregation. The size of such nanocubes varies at different time intervals, as revealed by UV-vis absorption spectra and TEM images (Fig. 19.8). Their high stability in water would be favorable for further chemical modification and systematic investigations.

Acknowledgments Vivian Wing-Wah Yam acknowledges support from the university development fund of the university of Hong Kong, and the university of Hong Kong foundation for educational development and research limited. This work has been supported by a central allocation vote (hku 2/05c) grant from the research grants council of Hong Kong special administrative region, people's republic of china.

References

Canales, F., Gimeno, M.C., Laguna, A., Jones, P.G., 1996, Aurophilicity at Sulfur Centers. Synthesis and Reactivity of the Complex [S(Au$_2$dppf)]; Formation of Polynuclear Sulfur-Centered Complexes. Crystal Structures of [S(Au$_2$dppf)] ·2CHCl$_3$, [(µ-Au$_2$dppf){S(Au$_2$dppf)}$_2$](OTf)$_2$ · 8CHCl$_3$, and [S(AuPPh$_2$Me)$_2$(Au$_2$dppf)](ClO$_4$)$_2$ 3CH$_2$Cl$_2$. *Journal of American Chemical Soci*ety, **118**, pp. 4839–4845.

Chan, C.W., Cheng L.K., Che, C.M., 1994a, Luminescent Donor-Acceptor Platinum(II) Complexes. *Coordination Chemistry Reviews*, **132**, pp. 87–97.

Chan, C.W., Wong, W.T., Che, C.M., 1994b, Gold(III) Photooxidants. Photophysical, Photochemical Properties, and Crystal Structure of a Luminescent Cyclometalated Gold(III) Complex of 2,9-Diphenyl-1,10-Phenanthroline. *Inorganic Chemistry*, **33**, pp. 1266–1272.

Corrigan, J.F., Fenske, D, 1997, New Copper Telluride Clusters by Light-Induced Tellurolate-Telluride Conversions. *Angewandte Chemie, International Edition*, pp. 1981–1983.

Deveson, A., Dehnen, S., Fenske, D., 1997, Syntheses and Structures of Four New Copper(I)–Selenium Clusters: Size Dependence of the Cluster on the Reaction Conditions. *Journal of Chemical Society, Dalton Transaction*, pp. 4491–4498.

Fredericks, S.M., Luong, J.C., Wrighton, M.S., 1979, Multiple Emissions from Rhenium(I) Complexes: Intraligand and Charge-Transfer Emission from Substituted Metal Carbonyl Cations. *Journal of American Chemical Society*, **101**, pp. 7415–1717.

Grohmann, A., Schmidbaur, H., in: Abel, E.W., Stone, F.G.A., Wilkinson, G. (Eds.), 1995, *Comprehensive Organometallic Chemistry II*, vol. 3 (Pergamon, Oxford).

Lo, W.Y., Lam, C.H., Yam, V.W.W., Zhu, N., Cheung, K.K., Fathallah, S., Messaoudi, S., Le Guennic, B., Kahlal, S., Halet, J.F., 2004, Synthesis, Photophysics, Electrochemistry, Theoretical, and Transient Absorption Studies of Luminescent Copper(I) and Silver(I) Diynyl Complexes. X-Ray Crystal Structures of [Cu$_3$(µ-dppm)$_3$(µ$_3$-η^1-C≡CC≡CPh)$_2$]PF$_6$ and [Cu$_3$(µ-dppm)$_3$(µ$_3$-η^1-C≡CC≡CH)$_2$]PF$_6$. *Journal of American Chemical Society, 126*, pp. 7300–7310.

Marsh, R.E., Miskowski, V.M., Schaefer, W.P., Gray, H.B., 1995, Electronic Spectroscopy of Chloro(terpyridine)platinum(II). *Inorganic Chemistry*, **34**, pp. 4591–4599.

McMillin D.R., Moore, J.J., 2002, Luminescence that Lasts from Pt(trpy)Cl$^+$ derivatives (trpy = 2, 2′; 6′, 2″-terpyridine). *Coordination Chemistry Reviews*, **229**, pp. 113–121.

Miskowski V.M., Houlding, V.H., 1989, Electronic Spectra and Photophysics of Platinum(I1) Complexes with α-Diimine Ligands. Solid-state Effects. 1. Monomers and Ligand π Dimers. *Inorganic Chemistry*, **28**, pp. 1529–1533.

Tang, W.S., Lu, X.X., Wong, K.M.C., Yam, V.W.W., 2005, Synthesis, Photophysics and Binding Studies of Platinum(II) Alkynyl Terpyridine Complexes with Crown Ether Pendant. Potential Luminescent Sensors for Metal Ions. *Journal of Materials Chemistry*, **15**, pp. 2714–2720.

Wong, K.M.C., Tang, W.S., Chu, B.W.K., Zhu, N., Yam, V.W.W., 2004, Synthesis, Photophysical Properties and Biomolecular Labeling Studies of Luminescent Platinum(II)-Terpyridyl Alkynyl Complexes. *Organometallics*, **23**, pp. 3459–3465.

Wong, K.M.C., Tang, W.S., Lu, X.X., Zhu, N., Yam, V.W.W., 2005a, Functionalized Platinum(II) Terpyridyl Alkynyl Complexes as Colorimetric and Luminescence pH Sensors. *Inorganic Chemistry*, **44**, pp.1492–1498.

Wong, K.M.C., Zhu, X., Hung, L.L., Zhu, N., Yam, V.W.W., Kwok, H.S., 2005b, A Novel Class of Phosphorescent Gold(III) Alkynyl-Based Organic Light Emitting Devices with Tunable Color. *Chemical Communications*, pp. 2906–2908.

Wrighton, M.S., Morse, D.L., 1974, Nature of the lowest excited state in tricarbonylchloro-1,10-phenanthrolinerhenium(I) and related complexes. *Journal of American Chemical Society*, **96**, pp. 998–1003.

Yam, V.W.W., 2001, Luminescent Carbon-Rich Rhenium(I) Complexes. *Chemical Communications*, pp. 789–796.

Yam, V.W.W., Chan, K.H.Y., Wong, K.M.C., Zhu, N., 2005a, Luminescent Platinum(II) Terpyridyl Complexes: Effect of Counter Ions on Solvent-Induced Aggregation and Color Changes. *Chemistry - A European Journal*, **11**, pp. 4535–4543.

Yam, V.W.W., Wong, K.M.C., Hung, L.L., Zhu, N., 2005b, Luminescent Gold(III) Alkynyl Complexes - Synthesis, Structural Characterization and Luminescent Properties. *Angewandte Chemie International Edition*, **44**, pp. 3107–3110.

Yam, V.W.W., Cheng, E.C.C., Zhou, Z.Y., 2000a, A Highly Soluble Luminescent Decanuclear Gold(I) Complexes with a Propeller-Shaped Structure. *Angewandte Chemie, International Edition*, **39**, pp. 1683–1685; *Angewandte Chemie*, 2000, pp. 1749–1751.

Yam, V.W.W., Cheng, E.C.C., Cheung, K.K., 1999, A Novel High Nuclearity Luminescent Gold(I) Sulfido Complex. *Angewandte Chemie, International Edition in English*, **38**, pp. 197–199; *Angewandte Chemie*, **111**, pp. 193–195.

Yam, V.W.W., Chong, S.H.F., Ko, C.C., Cheung, K.K., 2000b, Synthesis and Luminescence Behaviour of Rhenium(I) Triynyl Complexes. X-Ray Crystal Structures of [Re(CO)$_3$(^{t}Bu$_2$bpy)(C$\equiv$C-C$\equiv$C-C$\equiv$CPh)] and [Re(CO)$_3$(Me$_2$bpy)(C$\equiv$C-C$\equiv$C-C$\equiv$CSiMe$_3$)]. *Organometallics*, **19**, pp. 5092–5097.

Yam, V.W.W., Cheng, E.C.C., Zhu, N., 2001a, A Novel Polynuclear Gold-Sulfur Cube with an Unusually Large Stoke's Shift. *Angewandte Chemie, International Edition*, **40**, pp. 1763–1765.

Yam, V.W.W., Cheng, E.C.C., Zhu, N., 2001b, The First Luminescent Tetranuclear Copper(I) μ_4-Phosphinidene Complex. *Chemical Communications*, pp. 1028–1029.

Yam, V.W.W., Tang, R.P.L., Wong, K.M.C., Cheung, K.K., 2001c, Synthesis, Luminescence and Ion-Binding Studies of Platinum(II) Terpyridyl Acetylide Complexes. *Organometallics*, **20**, pp. 4476–4482.

Yam, V.W.W., Choi, S.W.K., Lai, T.F., Lee, W.K., 1993a, Syntheses, Crystal Structures and Photophysics of Organogold(III) Diimine Complexes. X-Ray Crystal Structures of [Au^{III}(bpy)(mes)$_2$]ClO$_4$ and [Au^{III}(phen)(CH$_2$SiMe$_3$)$_2$]ClO$_4$. *Journal of Chemical Society, Dalton Transactions*, pp. 1001–1002.

Yam, V.W.W., Lee, W.K., Lai, T.F., 1993b, Synthesis and Luminescent Properties of a Novel Tetranuclear Copper(I) Cluster containing a μ_4-Sulphur Moiety. X-Ray Crystal Structure of [Cu$_4$(μ-dppm)$_4$(μ_4-S)](PF$_6$)$_2$ • 2Me$_2$CO [dppm = bis(diphenylphosphino) methane]. *Journal of Chemical Society, Chemical Communications*, pp. 1571–1573.

Yam, V.W.W., Lee, W.K., Lai, T.F., 1993c, Synthesis, Spectroscopy and Electrochemistry of Trinuclear Copper(I) Acetylides. X-Ray Crystal Structure of [Cu$_3$(μ-Ph$_2$PCH$_2$PPh$_2$)$_3$(μ_3-η^1-C$\equiv$C^{t}Bu)(μ_3-Cl)]PF$_6$. *Organometallics*, **12**, pp. 2383–2387.

Yam, V.W.W., Chong, S.H.F., Cheung, K.K., 1998, Synthesis and Luminescence Behaviour of Rhenium(I) Diynyl Complexes Complexes. X-Ray Crystal Structures of [Re(CO)$_3$(^{t}Bu$_2$bpy)(C$\equiv$C-C$\equiv$CH)] and [Re(CO)$_3$(^{t}Bu$_2$bpy)(C$\equiv$C-C$\equiv$CPh)]. *Chemical Communications*, pp. 2121–2122.

Yam, V.W.W., Fung, W.K.M., Cheung, K.K., 1996a, Synthesis, Structure, Photo-physics, and Excited-State Redox Properties of the Novel Luminescent Tetranuclear Acetylidocopper(I) Complex [Cu$_4$(μ-dppm)$_4$(μ_4-η^1, η^2-C$\equiv$C-)][BF$_4$]$_2$. *Angewandte Chemie, International Edition in English*, **35**, pp. 1100–1102; *Angewandte Chemie*, **108**, pp. 1213–1215.

Yam, V.W.W., Lo, K.K.W., Cheung, K.K., 1996b, A Novel Luminescent μ_4-Selenido-Bridged Copper(I) Tetramer. *Inorganic Chemistry*, **35**, pp. 3459–3462.

Yam, V.W.W., Lo, K.K.W., Wang, C.R., Cheung, K.K., 1996c, The First Series of Luminescent (μ_4-Chalcogenido)silver(I) Clusters. *Inorganic Chemistry*, **35**, pp. 5116–5117.

Yam, V.W.W., Lau, V.C.Y., Cheung, K.K., 1996d, Luminescent Rhenium(I) Carbon Wires: Synthesis and Photophysics. X-Ray Crystal Structure of [Re(^{t}Bu$_2$bpy)(CO)$_3$ ($C \equiv C$-C$\equiv$C)Re(^{t}Bu$_2$bpy)(CO)$_3$]. *Organometallics*, **15**, pp. 1740–1744.

Yam, V.W.W., Fung, W.K.M., Wong, M.T., 1997, Synthesis, Photophysics, Electrochemistry and Excited-State Redox Properties of Trinuclear Copper(I) Acetylides with Bis(diphenylphosphino)-Alkyl and Aryl Amines as Bridging Ligands. *Organometallics*, **16**, pp. 1772–1778.

Yam, V.W.W., Lau, V.C.Y., Cheung, K.K., 1995, Synthesis and Photophysics of Luminescent Rhenium(I) Acetylides - Precursors for Organometallic Rigid-Rod Materials. X-Ray Crystal Structures of [Re(^{t}Bu$_2$bpy)(CO)$_3$(^{t}BuC$\equiv$C)] and [Re(^{t}Bu$_2$bpy)(CO)$_3$Cl]. *Organometallics*, **14**, pp. 2749–2753.

Yam, V.W.W., Li, B., Zhu, N., 2002a, Synthesis of Mesoporous Silicates with Controllable Pore Size Using Surfactant Ruthenium(II) Complexes that Display Lyotropic Mesomorphism as Templates. *Advanced Materials*, **14**, pp. 719–722.

Yam, V.W.W., Wong, K.M.C., Zhu, N, 2002b, Solvent-Induced Aggregation through Metal$\cdots$ Metal/$\pi\cdots\pi$ Interactions – Large Solvatochromism of Luminescent Organoplatinum(II) Ter-pyridyl Complexes. *Journal of American Chemical Society*, **124**, pp. 6506–6507.

Yam, V.W.W., Lo, K.K.W. 1997, Luminescent Tetranuclear Copper(I) and Silver(I) Chalcogenides. *Comments on Inorganic Chemistry*, **19**, pp. 209–229.

Yip, S.K., Cheng, E.C.C., Yuan, L.H., Zhu, N., Yam, V.W.W., 2004, Supramolecular Assembly of Luminescent Gold(I) Alkynylcalix[4]crown-6 Complexes Having Unprecedented Planar η^2, η^2-Coordination Gold(I) Centers. *Angewandte Chemie International Edition*, **43**, pp. 4954–4957.

Yu, D.B., Yam, V.W.W., 2004, Controlled Synthesis of Monodisperse Silver Nanocubes in Water. *Journal of American Chemical Society*, **126**, pp. 13200–13201.

Yu, D.B., Yam, V.W.W., 2005, Hydrothermal-Induced Assembly of Colloidal Silver Spheres into Various Nanoparticles on the Basis of HTAB-Modified Silver Mirror Reaction. *Journal of Physical Chemistry B*, **109**, pp. 5497–5503.

Chapter 20
First-Principles Method for Open Electronic Systems

Xiao Zheng and GuanHua Chen*

We prove the existence of the exact density-functional theory formalism for open electronic systems, and develop subsequently an exact time-dependent density-functional theory (TDDFT) formulation for the dynamic response. The TDDFT formulation depends in principle only on the electron density of the reduced system. Based on the nonequilibrium Green's function technique, it is expressed in the form of the equation of motion for the reduced single-electron density matrix, and this provides thus an efficient numerical approach to calculate the dynamic properties of open electronic systems. In the steady-state limit, the conventional first-principles nonequilibrium Green's function formulation for the current is recovered.

Density-functional theory (DFT) has been widely used as a research tool in condensed matter physics, chemistry, materials science, and nanoscience. The Hohenberg-Kohn theorem (1964) lays the foundation of DFT. The Kohn-Sham formalism (1965) provides the practical solution to calculate the ground state properties of electronic systems. Runge and Gross extended further DFT to calculate the time-dependent properties and hence the excited state properties of any electronic systems (Runge and Gross, 1984). The accuracy of DFT or TDDFT is determined by the exchange-correlation functional. If the exact exchange-correlation functional were known, the Kohn-Sham formalism would have provided the exact ground state properties, and the Runge-Gross extension, TDDFT, would have yielded the exact properties of excited states. Despite of their wide range of applications, DFT and TDDFT have been mostly limited to closed systems.

Fundamental progress has been made in the field of molecular electronics recently. DFT-based simulations on quantum transport through individual molecules attached to electrodes offer guidance for the design of practical devices (Lang and Avouris, 2000; Heurich et al., 2002; Wang and Luo, 2003). These simulations focus on the steady-state currents under the bias voltages. Two types of approaches have been adopted. One is the Lippmann-Schwinger formalism by Lang and coworkers (Lang, 1995). The other is the first-principles nonequilibrium Green's function technique (Taylor et al., 2001; Ke et al., 2004; Deng et al., 2004; Brandbyge et al., 2002; Xue et al., 2001). In both approaches the Kohn-Sham

---

* Department of Chemistry, The University of Hong Kong, Pokfulam Road, Hong Kong, China

Z. Tang, P. Sheng (eds), *Nano Science and Technology.*
© Springer 2008

Fock operator is taken as the effective single-electron model Hamiltonian, and the transmission coefficients are calculated within the noninteracting electron model. It is thus not clear whether the two approaches are rigorous. Recently Stefanucci and Almbladh (2004) derived an exact expression for time-dependent current in the framework of TDDFT. In the steady-current limit, their expression leads to the conventional first-principles nonequilibrium Green's function formalism if the TDDFT exchange-correlation functional is adopted. However, they did not provide a feasible numerical formulation for simulating the transient response of molecular electronic devices. In this communication, we present a rigorous first-principles formulation to calculate the dynamic properties of open electronic systems. We prove first a theorem that the electron density distribution of the reduced system determines all physical properties or processes of the entire system. The theorem lays down the foundation of the first-principles method for open systems. We present then the equation of motion (EOM) for nonequilibrium Green's functions (NEGF) in the framework of TDDFT. By introducing a new functional for the interaction between the reduced system and the environment, we develop further a reduced-single-electron-density-matrix-based TDDFT formulation. Finally, we derive an exact expression for the current which leads to the existing DFT-NEGF formula in the steady-state limit. This shows that the conventional DFT-NEGF formalism can be exact so long as the correct exchange-correlation functional is adopted. Both Hohenberg-Kohn theorem and Runge-Gross extension apply to isolated systems. Applying Hohenberg-Kohn-Sham's DFT and Runge-Gross's TDDFT to open systems requires in principle the knowledge of the electron density distribution of the total system which consists of the reduced system and the environment. This presents a major obstacle in simulating the dynamic processes of open systems. Our objective is to develop an exact DFT formulation for open systems. In fact, we are interested only in the physical properties and processes of the reduced system. The environment provides the boundary conditions and serves as the current source and energy sink. We thus concentrate on the reduced system.

Any electron density distribution function $\rho(\mathbf{r})$ of a real physical system is a real analytic function. We may treat nuclei as point charges, and this would only lead to non-analytic electron density at isolated points. In practical quantum mechanical simulations, analytic functions such as Gaussian functions and plane wave functions are adopted as basis sets, which results in analytic electron density distribution. Therefore, we conclude that any electron density functions of real systems are real analytic on connected physical space. Note that the isolated points at nuclei can be excluded for the moment from the physical space that we consider, so long the space is connected. Later we will come back to these isolated points, and show that their inclusion does not alter our conclusion. Based on this, we show below that for a real physical system the electron density distribution function on a subspace determines uniquely its values on the entire physical space. This is nothing but the analytic continuation of a real analytic function. The proof for the univariable real analytical functions can be found in textbooks, for instance, reference (Krantz and Parks, 2002). The extension to the multivariable real analytical functions is straightforward.

Lemma 1. The electron density distribution function $\rho(r)$ is real analytic on a connected physical space U. $W \subseteq U$ is a sub-space. If $\rho(r)$ is known for all $\rho(r) \in W$, $\rho(r)$ can be uniquely determined on entire U.

Proof. To facilitate our discussion, the following notations are introduced. Set $\mathbb{Z}^+ = \{0, 1, 2, \ldots\}$, and γ is an element of $(\mathbb{Z}^+)^3$, *i.e.*, $\gamma = (\gamma_1, \gamma_2, \gamma_3) \in (\mathbb{Z}^+)^3$. The displacement vector $\mathbf{r}$ is denoted by the three-dimensional variable $x = (x_1, x_2, x_3) \in U$. Denote that $\gamma! = \gamma_1! \gamma_2! \gamma_3!$, $x^\gamma = x_1^{\gamma_1} x_2^{\gamma_2} x_3^{\gamma_3}$, and $\frac{\partial^\gamma}{\partial x^\gamma} = \frac{\partial^{\gamma_1}}{\partial x_1^{\gamma_1}} \frac{\partial^{\gamma_2}}{\partial x_2^{\gamma_2}} \frac{\partial^{\gamma_3}}{\partial x_3^{\gamma_3}}$.

Suppose that another density distribution function $\rho'(x)$ is real analytic on U and equal to $\rho(x)$ for all $x \in W$. We have $\frac{\partial^\gamma \rho(x)}{\partial x^\gamma} = \frac{\partial^\gamma \rho'(x)}{\partial x^\gamma}$ for all $x \in W$ and $\gamma \in (\mathbb{Z}^+)^3$. Taking a point x_0 at or infinitely close to the boundary of W, we may expand $\rho(x)$ and $\rho(x')$ around x_0, *i.e.*, $\rho(x) = \sum_{\gamma \in (\mathbb{Z}^+)^3} \frac{1}{\gamma!} \left. \frac{\partial^\gamma \rho(x)}{\partial x^\gamma} \right|_{x_0} (x - x_0)^\gamma$ and $\rho'(x) = \sum_{\gamma \in (\mathbb{Z}^+)^3} \frac{1}{\gamma!} \left. \frac{\partial^\gamma \rho'(x)}{\partial x^\gamma} \right|_{x_0} (x - x_0)^\gamma$. Assuming that the convergence radii for the Taylor expansions of $\rho(x)$ and $\rho'(x)$ at x_0 are both larger than a positive finite real number b, we have thus $\rho(x) = \rho'(x)$ for all $x \in D_b(x_0) = \{x : |x - x_0| < b\}$ since $\left. \frac{\partial^\gamma \rho(x)}{\partial x^\gamma} \right|_{x_0} = \left. \frac{\partial^\gamma \rho'(x)}{\partial x^\gamma} \right|_{x_0}$. Therefore, the equality $\rho'(x) = \rho(x)$ has been expanded beyond W to include $D_b(x_0)$. Since U is connected the above procedure can be repeated until $\rho'(x) = \rho(x)$ for all $x \in U$.

We have thus proven that ρ can be uniquely determined on U once it is known on W, and are ready to prove the following theorem.

Theorem 1. Electron density function $\rho(r)$ for a subsystem of a connected real physical system determines uniquely all electronic properties of the entire system.

Proof. Assuming the physical space spanned by the subsystem and the connected real physical system are W and U, respectively. W is thus a sub-space of U, i.e., $W \subseteq U$. According to the above lemma, $\rho(r)$ on W determines uniquely their value on U, i.e., $\rho(r)$ of the subsystem determines $\rho(r)$ of the entire system.

Inclusion of isolated points, lines or planes where $\rho(r)$ is non-analytic into the connected physical space does not violate the theorem; so long $\rho(r)$ is continuous on these points, lines or planes. This can be shown clearly by performing analytical continuation of $\rho(r)$ infinitesimally close to them. Therefore, we conclude that $\rho(r)$ of any finite subsystem determines uniquely $\rho(r)$ of the entire physical system including the nuclear sites. Hohenberg-Kohn theorem and Runge-Gross extension state that the electron density distribution of a system determines uniquely all its electronic properties. Therefore, we conclude that $\rho(r)$ for a subsystem determines all the electronic properties of the real physical system.

The above theorem guarantees the existence of an exact DFT-type method for open systems. In principle, all we need to know is the electron density of the reduced system. The electron density distribution in the environment can be obtained by the analytic continuation of the electron density function at or near the boundary. In practice the reduced system must be chosen appropriately so that this analytic continuation procedure can be conveniently performed. For instance, Fig. 20.1 depicts one type of open systems, a molecular device. It consists of the reduced system or

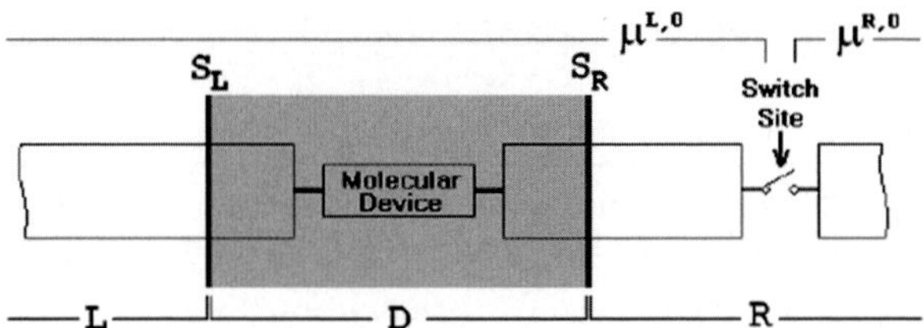

Fig. 20.1 Schematic representation of the experimental setup for quantum transport through a molecular device

device region D and the environment, the left and right leads L and R. Note that the reduced system D contains not only the molecular device itself, but also portions of the left and right electrodes, in this way the extension of the density function can be facilitated by the knowledge on the electron density distribution of the bulk electrodes.

The challenge is to develop a practical first-principles method. Taking the open system displayed in Fig. 20.1 as an example, we develop an exact DFT formalism for the open systems. To calculate the properties of a molecular device, we need only the electron density distribution in the device region. The influence of the electrodes can be determined by the electron density distribution in the device region. Within the TDDFT formalism, a closed nonlinear self-consistent EOM has been derived for the reduced single-electron density matrix $\sigma(t)$ of the entire system (Yam et al., 2003):

$$i\dot{\sigma}(t) = [h(t), \sigma(t)], \tag{20.1}$$

where $h(t)$ is the Kohn-Sham Fock matrix, and the square bracket on the right-hand side (RHS) denotes the commutator. The matrix element of σ is defined as

$$\sigma_{ij}(t) = \langle a_j^{\dagger}(t)a_i(t)\rangle, \tag{20.2}$$

where $a_i(t)$ and $a_j^{+}(t)$ are the Heisenberg annihilation and creation operators for atomic orbitals i and j at time t, respectively. Expanded in a real space basis set, the matrix representation of σ can be partitioned as

$$\sigma = \begin{bmatrix} \sigma_L & \sigma_{LD} & \sigma_{LR} \\ \sigma_{DL} & \sigma_D & \sigma_{DR} \\ \sigma_{RL} & \sigma_{RD} & \sigma_R \end{bmatrix}, \tag{20.3}$$

where σ_L, σ_R, and σ_D represent the diagonal blocks correspond to the left lead, the right lead and the device region, respectively, and σ_{LD}, σ_{RD}, σ_{LR}, σ_{DL}, σ_{DR} and σ_{RL} are the electron coherence between two different subsystems. The only block of finite dimensions is σ_D, The Kohn-Sham Fock matrix h can be partitioned in the same way with σ replaced by h in Eq. (20.3). Thus the EOM for σ_D which is of our central interest can be written down readily as

$$i\dot{\sigma}_D = [h_D, \sigma_D] + \sum_{\alpha=L.R} (h_{D\alpha}\sigma_{\alpha D} - \sigma_{D\alpha}h_{\alpha D}). \tag{20.4}$$

The reduced system D and the leads L/R are spanned by the atomic orbitals $\{l\}$ and the single-electron states $\{k_\alpha\}$, respectively. And hence Eq. (20.4) is equivalent to

$$i\dot{\sigma}_{nm} = \sum_{l\in D} (h_{nl}\sigma_{lm} - \sigma_{nl}h_{lm}) - i \sum_{\alpha=L,R} Q_{\alpha,nm}. \tag{20.5}$$

Here the $Q_{\alpha,nm}$ on the RHS is the dissipative term due to the lead α (L or R) whose expanded form is

$$Q_{\alpha,nm} = i \sum_{k\in\alpha} \left(h_{n,k_\alpha}\sigma k_\alpha, m - \sigma_{m,k_\alpha}h_{k_\alpha,m}\right), \tag{20.6}$$

where $h_{n,k\alpha}$ is the coupling matrix element between the atomic orbital n and the single-electron state k_α. $\sigma_{k\alpha,m}$ is precisely the lesser Green's function of identical time variables, i.e.,

$$\sigma_{k_{\alpha,m}}(t) = -i G^{<}_{k_{\alpha,m}}\left(t.t'\right)|_{t'=t}.$$

Based on the Keldysh formalism (1965), the analytic continuation rules of Langreth and Nordlander (1991) and the NEGF formulation developed by Jauho et al. (1994), $Q_{\alpha,nm}(t)$ can be calculated by

$$Q_{\alpha,nm}(t) = -\sum_{l\in D} \int_{-\infty}^{\infty} d\tau \left[G^{<}_{nl}(t,\tau) \sum_{\alpha,lm}^{\alpha} (\tau,t) + \right.$$
$$G^{r}_{nl}(t,\tau) \sum_{\alpha,lm}^{<} (\tau,t) - \sum_{\alpha,nl}^{<}(t,\tau) G^{\alpha}_{lm}(\tau,t)$$
$$\left. - \sum_{\alpha,nl}^{n}(t,\tau) G^{<}_{lm}(\tau,t) \right]. \tag{20.7}$$

where G^r, G^a and $G^<$ are the retarded, advanced and lesser Green's function for the reduced system D, and Σ^r, Σ^a, and $\Sigma^<$ are the retarded, advanced and lesser self-energies, respectively. The current through the interfaces S_L or S_R (see Fig. 20.1) can be expressed as

$$J_\alpha(t) = -\int_\alpha d\mathbf{r} \frac{\partial}{\partial t}\rho(\mathbf{r},t) = -\sum_{k\in\alpha} \frac{d}{dt}\sigma_{k_\alpha,k_\alpha}(t)$$
$$= i \sum_{l\in D}\sum_{k\in\alpha} \left(h_{k_\alpha,l}\sigma_{l,k_\alpha} - \sigma_{k_\alpha,l}h_{l,k_\alpha}\right)$$
$$= -\sum_{l\in D} Q_{\alpha,ll} = -\mathrm{tr}\left[Q_\alpha(t)\right]. \tag{20.8}$$

i.e., the current through S_α ($\alpha=L$ or R) is the negative of the trace of Q_α. Combining Eqs. (20.7) and (20.8), we recover the current expression (Eq. 20.8) of (Stefanucci and Almbladh, 2004).

At first glance Eq. (20.5) is not self-closed since the dissipative terms Q_α remain unsolved. According to the theorem we proved earlier, all physical quantities are explicit or implicit functionals of the electron density in D, $\rho_D(t)$. Q_α are thus also universal functionals of $\rho_D(t)$. Therefore, Eq. (20.5) can be recast into a formally closed form,

$$i\dot\sigma_D = [h_D\,[\rho_D\,(t)]\,,\sigma_D] - i\sum_{\alpha=L,R} Q_\alpha\,[\rho_D\,(t)]. \qquad (20.9)$$

Neglecting the second term on the RHS of Eq. (20.9) leads to the conventional TDDFT formulation in terms of reduced single-electron density matrix (Yam et al., 2003). The second term describes the dissipative processes where electrons enter and leave the region D. Besides the exchange-correlation functional, the additional universal density functional $Q_\alpha[\rho_D(t)]$ is introduced to account for the dissipative interaction between the reduced system and its environment. Equation (20.9) is thus the TDDFT formulation in terms of the reduced single-electron matrix for the open system. In the frozen DFT approach (Wesolowski and Warshel, 1993) an additional exchange-correlation functional term was introduced to account for the exchange-correlation interaction between the system and the environment. This additional term is included in $h_D(t)$ of Eq. (20.5). Yokojima et al. (2003) developed a dynamic mean-field theory for dissipative interacting many-electron systems. An EOM for the reduced single-electron density matrix was derived to simulate the excitation and nonradiative relaxation of a molecule embedded in a thermal bath. This is in analogy to our case although our environment is actually a fermion bath instead of a boson bath. The major difference is that the number of electrons in the reduced system is conserved in reference (Yokojima et al., 2003) while in our case it is not. Admittedly, $Q_\alpha[\rho_D(t)]$ can be an extremely complex functional. Progressive approximations are needed for the practical solution of Eq. (20.9).

To obtain the steady-state solution of Eqs. (20.5) or (20.9), we adopt a similar strategy as that of reference (Stefanucci and Almbladh, 2004). As $\tau \to \infty$, $\Gamma_{nm}^{k_\alpha}(t,\tau) \equiv h_{nk_\alpha}(t)h_{k_\alpha m}(\tau)$ becomes asymptotically time-independent, and Gs and Σs rely simply on the difference of the two time-variables (Stefanucci and Almbladh, 2004). The steady-state current can thus be expressed as

$$J_L\,(\infty) = -\,J_R\,(\infty) = -\sum_{n\in D} Q_{L,nn}\,(\infty)$$

$$= \int \left[f^L\,(\epsilon) - f^R\,(\epsilon)\right] T\,(\epsilon)\,d\epsilon, \qquad (20.10)$$

$$T(\epsilon) = 2\pi\eta_L\,(\epsilon)\,\eta_R\,(\epsilon)\times$$

$$\mathrm{tr}\left[G_D^r\,(\epsilon)\,\Gamma^R\,(\epsilon)\,G_D^\alpha\,(\epsilon)\,\Gamma^L\,(\epsilon)\right] \qquad (20.11)$$

Here $T(\varepsilon)$ is the transmission coefficient, $f^{\alpha}(\varepsilon)$ is the Fermi distribution function, and $\eta_{\alpha}(\varepsilon) = \sum_{k \in \alpha} \delta(\varepsilon - \varepsilon_k^{\alpha})$ is the density of states for the lead α (L or R). Equation (20.10) is exactly the Landauer formula (Datta, 1995; Landauer, 1970) in the DFT-NEGF formalism (Taylor et al., 2001; Ke et al., 2004). The only difference is that Eq. (20.10) is derived within the TDDFT formalism in our case while it is evaluated within the DFT framework in the case of the DFT-NEGF formulation (Taylor et al., 2001; Ke et al., 2004). In other words, *the DFT-NEGF formalism can be exact so long as the correct exchange-correlation functional is used!* This is not surprising, and is simply a consequence of that (*i*) DFT and TDDFT can yield the exact electron density and (*ii*) the current is the time derivative of the total charge.

Just as the exchange-correlation functional, the exact functional form of Q_α on density is rather difficult to derive. Various approximated expressions have been adopted for the DFT exchange-correlation functional in practical implementations. Similar strategy can be employed for Q_α One such scheme is the adiabatic approximation, which implies that the dissipation functional Q_α depends instantaneously on the electron density of the reduced system D, i.e.,

$$Q_\alpha \left[\rho_D(t) \right] \approx Q_\alpha^{AD} \left[\rho_D^t \right], \tag{20.12}$$

where Q_α is a functional of the density function $\rho_D(t)$ over both time and space and Q_α^{AD} is a functional of $\rho_D(t)$ over space only. For those $\rho_D(t)$ corresponding to the steady-state solutions of Eq. (20.9), the exact functional values of $Q_\alpha^{AD}[\rho_D(t)]$ can be acquired by the TDDFT-NEGF formalism with the correct exchange-correlation functional being adopted, i.e.,

$$Q_\alpha^{AD} \left[\rho_D^t \right] \big|_{\rho_D^t \in \{\rho_D^{SS}\}} = Q_\alpha^{TDDFT-NEGF}[\rho_D^t], \tag{20.13}$$

where $\{\rho_D^{SS}\}$ denotes the set of all accessible steady-state density functions on the reduced system D. The explicit functional form of Q_α^{AD} can thus be obtained by fitting to the data available from Eq. (20.13) and then smoothly extrapolating to $\rho_D(t)$ not in $\{\rho_D^{SS}\}$. In this way Q_α becomes an explicit functional of ρ_D and the Eq. (20.9) spontaneously leads to the same correct steady-state solution as that offered by the TDDFT-NEGF formalism.

Another scheme is the wide-band limit (WBL) approximation (Jauho et al., 1994), which consists of a series of approximations imposed on the leads: (*i*) their band-widths are assumed to be infinitely large, (*ii*) their linewidths $\Lambda_k^{\alpha}(t, \tau)$ defined by $\pi \eta_{\alpha}(\varepsilon_k^{\alpha}) \Gamma^{k\alpha}(t, \tau)$ are regarded as energy independent, i.e.,

$$\Lambda_k^{\alpha}(t, \tau) \approx \Lambda^{\alpha}(t, \tau) \approx \Lambda^{\alpha},$$

and (*iii*) the energy shifts are taken as level independent, i.e., $\delta \varepsilon_k^{\alpha}(t) \approx \delta \varepsilon^{\alpha}(t) \approx \delta \varepsilon^{\alpha}$ for L or R. The physical essence of the transport problem is captured under these reasonable hypotheses [18]. In the practical implementation, the effects of the specific electronic structures of the leads can be captured by enlarging the device region to include enough portions of the electrodes. Following the procedures similar to those in reference (Jauho et al., 1994), we obtain that

$$Q_\alpha = P^\alpha(t) + \left[P^\alpha(t)\right]^\dagger + \left\{\Lambda^\alpha \cdot \sigma_D\right\}, \qquad (20.14)$$

where the curly bracket on the RHS denotes the anticommutator. Taking $t = 0$ as the switch-on instant, $P^\alpha(t)$ can be written down as

$$P^\alpha(t) = \frac{2i}{\pi} U^{(-)}(t) \left\{ \int_0^t d\tau \frac{e^{i\delta\epsilon^\alpha(t-\tau)}}{t-\tau} U^{(+)}(\tau) + \int_{-\infty}^0 d\tau \frac{i e^{i\delta\epsilon^\alpha \tau}}{t-\tau} G_D^{\tau,0}(-\tau) \right\} \Lambda^\alpha - 2\Lambda^\alpha. \qquad (20.15)$$

where G_D^r is the retarded Green's function of D before the switch-on instant. The propagators for the reduced system D are defined as

$$U^{(\pm)}(t) = \exp\left\{ \pm i \int_0^t h_D(\tau)\, d\tau \pm \sum_{\alpha=L,R} \Lambda^\alpha t \right\} \qquad (20.16)$$

Equations (20.14) $\sim$ (20.16) constitute the WBL formulation of the TDDFT-NEGF formalism where Q_α exhibit an implicit dependence on $\rho_D(t)$.

To summarize, we have proven the existence of the exact TDDFT formalism for the open electronic systems, and have proposed a TDDFT-NEGF formulation to calculate the quantum transport properties of molecular devices. Since TDDFT results in formally exact density distribution, the TDDFT-NEGF formulation is in principle an exact theory to evaluate the transient and steady-state currents. In particular, the TDDFT-NEGF expression for the steady-state current has the exact same form as that of the conventional DFT-NEGF formalism (Taylor et al., 2001; Ke et al., 2004; Deng et al., 2004; Brandbyge et al., 2002; Xue et al., 2001), and this provides rigorous theoretical foundation for the existing DFT-based methodologies (Lang, 1995; Taylor et al., 2001; Ke et al., 2004; Deng et al., 2004; Brandbyge et al., 2002; Xue et al., 2001) calculating the steady currents through molecular devices. In addition to the conventional exchange-correlation functional, a new density functional is introduced to account for the dissipative interaction between the reduced system and the environment. In the WBL approximation, the new functional can be expressed in a relatively simple form which depends implicitly on the electron density of the reduced system. Since the basic variable in our formulation is the reduce single-electron density matrix, the linear-scaling techniques such as that of reference (Yam et al., 2003) can be adopted to further speed up the computation.

Acknowledgments Authors would thank Hong Guo, Jiang-Hua Lu, Jian Wang, Arieh Warshel and Weitao Yang for stimulating discussions. Support from The Hong Kong Research Grant Council (HKU 7010/03P) and the committee for Research and Conference Grants (CRCG) of The University of Hong Kong is gratefully acknowledged.

References

Brandbyge M. et al., 2002, *Phys. Rev.* B **65**, p. 165401.
Datta S., 1995, *Electronic Transport in Mesoscopic Systems* (Cambridge University Press).
Deng W.-Q., Muller R. P. and Goddard W. A. III, 2004, *J. Am. Chem. Soc.* **126**, p. 13563.
Heurich J., Cuevas J. C., Wenzel W. and Schön G., 2002, *Phys. Rev. Lett.* **88**, p. 256803.
Hohenberg P. and Kohn W., 1964, *Phys. Rev.* **136**, p. 64.
Jauho A.-P., Wingreen N. S. and Meir Y., 1994, *Phys. Rev.* B **50**, p. 5528.
Ke S.-H., Baranger H. U. and Yang W., 2004, *J. Am. Chem. Soc.* **126**, p. 15897.
Keldysh L. V., 1965, *JETP* **20**, p. 1018.
Kohn W. and Sham L. J., 1965, *Phys. Rev.* **140**, p. 1133.
Krantz S. G. and Parks H. R., 2002, *A Primer of Real Analytic Functions* (Birkhauser, Boston).
Landauer R., 1970, *Philos. Mag.* **21**, p. 863.
Lang N. D., 1995, *Phys. Rev.* B **52**, p. 5335.
Lang N. D. and Avouris Ph., 2000, *Phys. Rev. Lett.* **84**, p. 358.
Langreth D. C. and Nordlander P., 1991, *Phys. Rev.* B **43**, 2541.
Runge E. and Gross E. K. U., 1984, *Phys. Rev. Lett.* **52**, p. 997.
Stefanucci G. and Almbladh C.-O., 2004, *Europhys. Lett.* **67** (1), p. 14.
Taylor J., Guo H. and Wang J., 2001, *Phys. Rev.* B. **63**, p. 245407.
Wang C.-K. and Luo Y., 2003, *J. Chem. Phys.* **119**, p. 4923.
Wesolowski T. A. and Warshel A., 1993, *J. Phys. Chem.* **97**, p. 8050.
Xue Y., Datta S. and Ratner M. A., 2001, *J. Chem. Phys.* **115**, p. 4292.
Yam C. Y., Yokojima S. and Chen G. H., 2003, *J. Chem. Phys.* **119**, p. 8794; *Phys. Rev.* B **68**, p. 153105.
Yokojima S., Chen G. H., Xu R. and Yan Y., 2003, *Chem. Phys. Lett.* **369**, p. 495; *J. Comp. Chem.* **24**, p. 2083.

Author Index

Beling, C. D., 119

Chan, W. K., 119, 187
Chen, C. D., 211
Chen, C. H., 181
Chen, F. R., 141
Chen, G. H., 237
Chen, Y. T., 211
Cheng, H. M., 29, 41, 151
Cheng, I. C., 211
Cheng, K. W., 187
Cheung, C. K., 119
Chu, C. J., 211
Chu, M. W., 181
Cong, H. T., 151
Cui, X. D., 83

Datta, S., 59
Ding, L., 83
Djurišić, A. B., 119

Endo, M., 3, 9

Fung, S., 119

Ge, W. K., 83
Geng, H. Z., 15

He, H. T., 83
Hou, J. G., 203

Kai, J. J., 141

Lee, T. C., 141
Lee, Y. H.15
Lei, S. B., 135
Leung, Y. H., 119
Li, F., 41
Li, I. L., 49
Li, Z. M., 49
Lin, H. Y., 211
Lin, J. J., 141

Lin, K. J., 141
Lin, M. C., 211
Lin, Y. H., 141
Ling, C. C., 119
Liu, C., 41, 151
Liu, C. J., 211

Man, K. Y., 187

Qian, T. Z., 99

Ren, W. C., 29

Shen, S. Q., 83
Sheng, P., 99
Shieh, D. B., 211

Tam, K. H., 119
Tang, D. M., 151
Tang, Y. B., 151
Tang, Z. K., 49, 193
Tsai, L. C., 211
Tse, C. W., 187

Wang, C., 135
Wang, J. N., 83
Wang, X. P., 99
Wong, K. M. C., 219
Wu, C. S., 211
Wu, H. C., 211
Wu, Y. N., 211
Wu, Y. P., 211

Yam, V. W. W., 219
Yang, C. L., 83
Ye, J. T., 49, 193

Zhai, J. P., 49
Zhang, F. C., 83
Zhang, H. L., 41
Zheng, X., 237

Subject Index

AlN, 149–174
Aluminophosphate, 50

Ballistic transport, 61, 62

Capacitor, 79
Carbon nanotubes, 3
   anode material, 5, 41, 42, 44, 45, 47
   band gap, 20, 21
   capacitor, 5, 6, 29
   carbon nanotube-filled nylon, 11
   catalytically grown, 9
   conductance, 23, 24
   conductivity, 15, 20–25
   double-walled, 10
   electrode material, 5, 41, 44, 45, 47
   energy storage, 5, 11, 12, 41, 46
   field emission, 11, 12, 16
   large scale production, 5, 9, 10
   lithium ion battery, 5, 41, 44, 47
   material quality factor, 16, 20,
      22–25
   microcatheter, 11, 12
   multi-walled, 4, 9, 10, 29
   radial breathing mode, 19, 20, 22, 31–34,
      36, 38, 54, 55
   Raman, 16, 19, 20, 22, 25, 29–38, 45, 46,
      50–52, 54, 55
   resistance, 15–17, 23, 24
   semiconducting, 20–23, 34, 35
   SEM image, 4, 17, 25, 42, 43, 46
   single-walled (SWCNTs), 4, 9, 10, 16, 29,
      31, 33–37, 49, 50, 55
   solid electrolyte, 41
   TEM image, 5, 10, 17, 31, 33
   Transmittance, 15–17, 23–25
   transparent conducting films, 15
Circular photogalvanic effect, 83–89, 95
Conductance, 62–66, 72, 73, 76, 89, 204,
   209, 211

Conductivity, 62, 63, 140, 144, 145, 149, 150
CoPc, 205–207

Diffusive transport, 62, 109
DNA molecule, 209, 212–215

EELS, 152, 166, 170, 179–184
Electrochemical potential, 63, 64, 67, 68, 75
Electron gas, 83–86, 89–92, 95, 96, 179–181
Electronic transport, 139, 146
Electron spin, 83, 84, 86, 87
Energy dissipation, 60, 69, 99, 100, 103–107,
   114, 115
Equilibrium, 60, 63, 67–69, 73, 83, 105, 107,
   191, 235, 236

Field emission, 123, 141, 149, 151, 166–170,
   174, 182

Gold sulfido, 219, 220

Hydrodynamics, 99, 100, 110, 114
Hydrothermal Growth, 49, 50, 117, 118

Immiscible fluid, 99, 101, 102
InGaAs, 83, 84, 94, 96
Iodine molecular chain, 191, 192

Kerr rotation, 84, 89–91, 95

Mass spectra, 52–54
Maxwell's demon, 59–62, 71, 77

Nanorod, 117–119, 122–128, 151–155,
   167, 174
Nitride, 149–152, 174

Open electronic system, 235, 236, 242

Photoluminescence, 118–120, 122, 123, 152,
    169, 223
Photovoltaic, 185, 187–189
Platinum terpyridyl, 224–226
Pyrolysis reaction, 54

Quantum manipulation, 201, 202

Raman, 119, 120, 192–197
Resistance, 61, 85, 139–147, 151, 169, 185,
    188, 209, 213, 214
Rhenium diimine, 223
RuO$_2$, 140–143, 147

Semiconducting, 143, 149

SEM image, 124, 142, 146, 153, 155, 159–161,
    163–165, 167, 170–172, 211, 213
Silicon nanowire, 209–214
Spin-galvanic effect, 83
Spin photocurrent, 83–89, 95
STM, 133–137, 201–207

TEM image, 125, 153, 156, 160–166, 170,
    173, 231, 232
Tetranuclear copper, 218–222
Trinuclear copper, 221

Zeolite, 49, 53, 191–193, 197
Zinc oxide, 117, 123–128